Cambridge Elements

Elements in Metaphysics
edited by
Tuomas E. Tahko
University of Bristol

TELEOLOGY

Matthew Tugby
Durham University

Shaftesbury Road, Cambridge CB2 8EA, United Kingdom

One Liberty Plaza, 20th Floor, New York, NY 10006, USA

477 Williamstown Road, Port Melbourne, VIC 3207, Australia

314–321, 3rd Floor, Plot 3, Splendor Forum, Jasola District Centre, New Delhi – 110025, India

103 Penang Road, #05–06/07, Visioncrest Commercial, Singapore 238467

Cambridge University Press is part of Cambridge University Press & Assessment, a department of the University of Cambridge.

We share the University's mission to contribute to society through the pursuit of education, learning and research at the highest international levels of excellence.

www.cambridge.org
Information on this title: www.cambridge.org/9781009500326

DOI: 10.1017/9781009257404

© Matthew Tugby 2024

This publication is in copyright. Subject to statutory exception and to the provisions of relevant collective licensing agreements, no reproduction of any part may take place without the written permission of Cambridge University Press & Assessment.

When citing this work, please include a reference to the DOI 10.1017/9781009257404

First published 2024

A catalogue record for this publication is available from the British Library.

ISBN 978-1-009-50032-6 Hardback
ISBN 978-1-009-25739-8 Paperback
ISSN 2633-9862 (online)
ISSN 2633-9854 (print)

Cambridge University Press & Assessment has no responsibility for the persistence or accuracy of URLs for external or third-party internet websites referred to in this publication and does not guarantee that any content on such websites is, or will remain, accurate or appropriate.

Teleology

Elements in Metaphysics

DOI: 10.1017/9781009257404
First published online: June 2024

Matthew Tugby
Durham University

Author for correspondence: Matthew Tugby, matthew.tugby@durham.ac.uk

Abstract: Teleology is about functions, ends, and goals in nature. This Element offers a philosophical examination of these phenomena and aims to reinstate teleology as a core part of the metaphysics of science. It starts with a critical analysis of three theories of function and argues that functions ultimately depend on goals. A metaphysical investigation of goal-directedness is then undertaken. After arguing against reductive approaches to goal-directedness, the Element develops a new theory which grounds many cases of goal-directedness in the metaphysics of powers. According to this theory, teleological properties are genuine, irreducible features of the world.

Keywords: Teleology, functions, goals, ends, powers.

© Matthew Tugby 2024

ISBNs: 9781009500326 (HB), 9781009257398 (PB), 9781009257404 (OC)
ISSNs: 2633-9862 (online), 2633-9854 (print)

Contents

1	Introduction	1
2	Functions	7
3	Goals	25
4	Powers	41
	References	64

1 Introduction

You, dear reader, are a goal-directed system. Arguably, the snake plants and thermostats in my home are goal-directed systems too. On a much larger scale, the water and rock cycles that you learnt about in science classes at school might well be goal-directed systems. If these are all goal-directed systems, it is appropriate to ascribe functions to their parts. For example, the function of a thermostat's bimetallic strip is to gauge temperature.

By describing these examples in this way, we appear to be ascribing *teleological* properties to such systems. The term 'teleology' is derived from 'telos', the Greek word for 'end' or 'goal'. Teleology as a discipline is thus the study of ends and goals, as well as the related concepts of functions, aims, and purposes. What all these phenomena have in common is that they involve means–end relations of some sort. For instance, a goal-directed system acts in order to bring about a certain end (its goal); something with a function acts in order to contribute to the end (or goal) of a system; and something with a purpose typically acts in such a way as to achieve that purpose. It is also plausible that these teleological concepts are interrelated in various ways. For example, according to a theory explored in Section 2, the concepts of function and goal are intimately connected: something performs a function precisely when it contributes towards a goal of a system.

Teleology is not only of interest to metaphysicians, and I hope this contribution to the Elements series will encourage readers to engage with a range of theoretical themes across philosophy, science, and technology. Moreover, my aim is not only to explain teleology as a discipline, but also to provide my novel take on how teleology arises in the natural world. It is important to understand teleology because teleological concepts are employed by many explanations that are offered both in everyday contexts[1] as well as in the special sciences. Biological examples are often used in discussions of teleology, and I will often employ such examples in this Element. However, talk of functions is also common in (for instance) biochemistry, medicine, psychology, the social sciences and technology, as well as the emerging area of artificial intelligence (AI). Functional explanations are widespread in many of these areas, and many scientists regard such explanations as indispensable (Nagel 1979: 276). One hallmark of many functional explanations is that they are forward-looking: they explain the presence, character, or activity of an item by reference to some possible future outcome that the item tends to bring about. A mundane example

---

[1] Recent research in experimental philosophy suggests that our 'folk' intuitions about the world are thoroughly teleological (Kertész and Kodaj 2023; Rose and Schaffer 2017; Rose, Schaffer, and Tobia 2020).

of such an explanation is that a knife is sharp because the knife's function is to cut; but many functional explanations are more surprising than this and reflect important scientific discoveries. On the front page of the module guide for a metaphysics course on which I was once a teaching assistant, there was a picture of a donkey curling up its top lip. Amusingly, it appeared that the donkey was laughing or talking – a reference, I suspect, to David Lewis's famous example of donkeys talking in distant possible worlds. I always thought the donkey's pose was accidental, but later I learnt that the function of this lip movement in donkeys is to expose the vomeronasal organ, which has the further function of sensing pheromones in the air.[2] Importantly, because these functions contribute to the survival of donkeys in various respects, those functions also help to explain why these types of lip and organ *are there in the first place*. Thus, many philosophers of science (e.g., Garson 2017, 2019a, 2019b; Mitchell 1993; Neander 1991) have emphasized that many, or even all, biological functions have an important historical or 'backward-looking' explanatory role. We shall explore the details in Section 2, where we discuss the selectionist, causal role and goal-contribution conceptions of function. For current purposes, the important point is just that discoveries of various functional explanations are often startling, illuminating, and significant.

Now, the term 'function' does not always occur explicitly in the teleological explanations offered in biology and other sciences, as when we say, 'bees huddle together for the sake of keeping warm', or 'the bee wiggles its body in order to show other bees where to find pollen'. But locutions such as 'in order to', 'for the sake of', or 'for the purpose of' are nevertheless teleological insofar as they express a means–end relation, just like the concepts of goal, aim, and purpose. This is not to say, however, that all these terms are interchangeable: for instance, I take it that when putting forward various functional explanations, scientists are not implying that all goal-directed systems and subsystems are purposeful agents.

In the course of the twentieth century, functional explanations became more prominent than ever. In the 1960s, for example, biologists addressed questions about the functions of the thymus in vertebrates, and these investigations, among others, led to important work in immunology. Details aside, it turns out that thymus-derived 'T' cells play crucial roles in the production of lymphocytes, which in turn control cellular immunity (Miller 1961, 1971). This research, along with work on the functions of 'B' (bursa) cells, contributed to significant medical research programmes on cancer, autoimmune diseases, and organ transplant rejection (Schaffner 1993: 84).

---

[2] I am grateful to Andrea Komkov for explaining this example to me.

Perhaps the best-known development in the modern life sciences is the discovery of DNA; and again, it is difficult to explain the biological characteristics of DNA without invoking teleological concepts. For one thing, the functions of DNA contribute to the survival of a species, and therefore explain why DNA is there in the first place. Moreover, it seems to be in the very nature of DNA to encode information about future outcomes for an organism, and thus DNA appears to be inherently end-directed. As Feser puts it:

> in characterizing the DNA of bears, we take it to be relevant to note that it causes them to be furry and grow to a large size, but not that it also thereby causes them to be good mascots for football teams. The genetic information in bear DNA inherently 'points to' or is 'directed at' the first outcome, but not the second. (2009: 47)[3]

It may come as a surprise to students of philosophy and natural science that teleological concepts are deployed in so many branches of science. In the undergraduate classroom we are often told that natural teleology and the associated idea of final causation were eliminated from science and philosophy by Descartes and others during the early-modern scientific revolution. Teleology is thus often regarded as a bygone relic of ancient Aristotelian science. And yet, as noted earlier, many who work in the special sciences take teleological language and functional explanations to be indispensable.[4] Alongside the rise of modern science, over the last half-century there has also been a flourishing *metaphysical* debate about the nature of functions, giving rise to a philosophical cottage industry. This debate considers functions from a wide variety of scientific domains, and not just biology. So where does all this leave the early-modern rejection of teleology?

Here is one answer: when recent philosophers have said that modern science can do without teleology, what they typically meant is that talk of functions and goals in science can, *in principle*, be reduced to talk involving only non-teleological mechanistic concepts. Ernest Nagel is one early influential advocate of this reductive mechanistic approach, and we shall examine his account of goal-directed systems in Section 3. If Nagel is right, then the world is merely teleological with a small 't': scientists are *perfectly entitled* to employ teleological language, but such language is, in principle, dispensable.

---

[3] This is one way of thinking about the teleological nature of DNA, at any rate. Some caution is needed with such examples because, as a referee has helpfully pointed out, the literature on genetic functions is quite varied. For some different perspectives, see e.g., Doolittle 2013, Germain et al. 2014, and Bellazzi 2022.

[4] McDonough (2020b) also argues that the alleged rejection of teleology by early modern scientists and philosophers has been exaggerated by many historians. According to some recent scholars, even Spinoza is committed to teleology in some form (Sangiacomo 2015).

One reason to be attracted to the Nagelian reductive programme is that teleology is traditionally associated with the Aristotelian doctrine of the four causes and its underlying metaphysics of substantial forms, and many of us are sceptical about various aspects of Aristotelianism.[5] However, a point sometimes overlooked is that we can accept that there are teleological properties in the world without accepting *all* aspects of Aristotle's metaphysics. Aristotle, for example, argued that the world is imbued with inherent causal powers, and according to many Aristotelian scholars all powers – and not just those of goal-directed systems – are thoroughly teleological, that is, end-directed. As Witt puts it, for Aristotle 'A dormant power is intrinsically dependent upon, and teleologically directed toward, activity, or actuality, and that is the character of its being: it exists potentially' (2008: 130; see also Witt 2003). Importantly, one can be a realist about such power properties without accepting other aspects of Aristotelian physics or metaphysics, such as the controversial doctrine of substantial forms.

Realism about powers has attracted significant support in recent philosophy of science. For instance, it has been argued that even in physics it is plausible to think that what scientists are often doing is uncovering the fundamental dispositional properties or 'powers' of entities (e.g., Bird 2007; Cartwright 1992, 1999, 2019; Cartwright and Pemberton 2013; Corry 2019; Ellis 2001; Ellis and Lierse 1994; Kistler 2006; Mumford 2006). For example, the property of mass is often defined in terms of its gravitational and inertial powers. And importantly for our purposes, power properties are arguably needed to make sense of end-directedness in nature—in a way that reductive analyses of teleological statements fail to do. We explore these issues in Section 4. Our ultimate aim is to cast realism about teleology is a new positive light and encourage new work on it. If our proposals are correct, then the powers metaphysics might well rehabilitate a worldview in which reality is teleological with a big 'T'.

Another traditional complaint about teleology is that it implies something like backwards causation. The worry is that if the current presence, character, or activity of an item is being explained by some future outcome, then that outcome must be reaching out from the future in such a way as to have a causal influence on current states of affairs.[6] While some philosophers do entertain the metaphysical possibility of backwards causation, few would regard it as an actual phenomenon, never mind a widespread one. However, the backwards causation worry arguably

---

[5] In brief, substantial forms are kind universals like *proton* or *donkey*, which are instantiated essentially by their members and help to explain and unify the attributes of individuals. For a detailed discussion of this idea, see Oderberg 2007, and for discussion of a serious problem, see Alvarado and Tugby 2021.

[6] Gunnar Babcock tells me that this (misguided) objection probably originates in Spinoza's *Ethics* (2018/1677).

rests on a misunderstanding. For one thing, teleological explanations do not entail that the relevant future occurrence ever comes about. Clearly, a knife can have the function to cut even if it never has the occasion to cut anything. But if a cutting occurrence never comes to pass, there can be no question of backwards causation occurring. Arguably, then, function statements are at least partly about particular here-and-now capabilities of an item which may or may not be manifested. This idea will come out clearly in Section 4, where we discuss the idea that dispositions or powers are real here-and-now properties of things. This by no means resolves all the problems, of course, because the possibility of unmanifested functions leads to difficult questions about the nature of goal-directedness, which we raise in Sections 3 and 4. However, those difficulties are not difficulties concerning backwards causation.

When discussing theories of function, I shall assume that an adequate account should satisfy some interrelated desiderata. First and foremost, the theory should shed light on how true judgements about functions in science are grounded by aspects of the world around us (see Forber 2020: 260 for a similar thought). Call this the truthmaking desideratum ('TD'):

TD: An adequate theory of function should specify worldly truthmakers for true function claims.

Secondly, and relatedly, the theory should aim at extensional adequacy, so that it is at least roughly consistent with how teleological concepts are deployed in scientific practice and everyday discourse. Call this the extensional adequacy desideratum ('EAD').

EAD: An adequate theory of function should (at least roughly) preserve the extension of function concepts as they are ordinarily used.

Finally, I follow Garson (2016: Ch. 1.2) and Forber (2020: 261–262) in thinking that an adequate theory of functions should accommodate the normative and explanatory dimensions of function statements. That is, the theory should shed light on the various explanatory roles of function statements in science and also underwrite the normative distinctions between normal and accidental function, and between non-function and malfunction. Call these the explanatory desideratum ('ED') and the normative desideratum ('ND'), respectively.

ED: An adequate theory of function should accommodate, and shed light on, the explanatory roles of function statements in science.

ND: An adequate theory of function should accommodate, and shed light on, the distinctions between normal and accidental function, and between non-function and malfunction.

In the following sections, I discuss these desiderata further and make a case for thinking that a goal-contribution theory of functions (suitably understood) can satisfy them. Where further work is required, which will be inevitable at some points, this will be indicated.

I have had to make some difficult decisions about what to include and what to exclude in this Element. Teleology has a long history, going back at least to Plato and Aristotle, via (among others) Aquinas and the Scholastics in the medieval period, and major figures in early modern philosophy such as Hegel (2010/1816), Kant (2007/1790), and Schelling (2000/1800). Some historical theories of teleology are naturalistic, while others have been grounded in theism. Outside of metaphysics, teleology has also played a prominent role in disciplines including aesthetics, AI, cosmology, environmental philosophy, epistemology and perception, ethics, linguistics, metaethics, philosophy of action, philosophy of language, philosophy of mind, and philosophy of religion. This is not a long Element, and it would be unrealistic here to attempt even a brief survey of work on teleology in such a wide variety of intellectual periods, traditions, and disciplines. Moreover, excellent historical surveys have been provided elsewhere in the philosophical literature.[7] I have therefore confined this Element to work on natural teleology in recent metaphysics of science, which focuses mainly on the concepts of functions, goals, and powers. This approach is (moderately) naturalistic and therefore I shall, for example, have little to say about theistic accounts of teleology or teleological arguments for creationism, in which teleology is ultimately grounded extrinsically in some divine or supernatural power.[8] Nor shall I have much to say about alleged evaluative aspects of teleology. For some neo-Aristotelians, teleology is bound up with the evaluative notion of goodness, insofar as the good is that which fulfils a thing's natural end (for discussion, see e.g., Bedau 1992a; McLaughlin 2001: Ch. 9; Oderberg 2020; Page 2021; Sorabji 1964). The question of whether function statements entail a value judgement is, however, a difficult one that rests upon complex issues in value theory and metaethics. I have therefore not attempted to address the evaluative dimensions of teleology (or lack thereof) in this Element.

Here, then, is a roadmap of what follows: Section 2 offers an overview of the modern debate about functions, which first began to take shape in the 1970s.

---

[7] For further discussion of historical work on teleology in various intellectual periods, see for example, Feser 2014, 2019: Ch. 6, McDonough 2020a, and Ransome Johnson 2005.

[8] This is not to say that the theories discussed here are incompatible with theology. Indeed, many theistic approaches are inspired by aspects of Aristotelianism and realism about powers. For example, see Oliver 2013, Page 2015, and Schmid 2011 for recent discussions of theistic teleology in the Thomistic tradition.

Here we discuss the selectionist, causal role, and goal-contribution approaches. We make a preliminary case for the goal-contribution theory and show that it is consistent with many recent theories of function in the philosophy of biology and other scientific domains. Although the goal-contribution theory accommodates functions that are not naturally selected, it can readily acknowledge the importance of selected functions in biology and the role they play in explaining the very existence of function bearers. Section 3 then focuses on the concepts of goals and goal-directedness. It critically examines the reductive cybernetic theory of goal-directed systems that became popular in the middle part of the twentieth century. Section 4 then develops a non-reductive realist theory of powers and end-directedness, with a view to shed light on many cases of goal-directedness.

Each of the Sections in this Element can to some extent be read in isolation. However, they are also related in important ways, and the ordering of the Sections is not accidental. The goal-contribution theory of function, discussed in Section 2, depends on the concepts of goals and goal-directedness. Hence, it is necessary to delve deeper into those concepts in Section 3. However, we shall see that some traditional attempts to analyse goal-directedness face serious problems. This leads us to Section 4, where we consider whether a realist metaphysics of powers can shed light on goal-directedness, and end-directedness more generally. If the powers theory can be shown to do this, then we will be able to bring to light an important new benefit of the powers metaphysics. In the course of the discussion, it is also our aim to open up new and fruitful avenues of debate within the metaphysics of teleology.

2 Functions

2.1 The Concept of Function

As explained in the introductory section, teleology, as I shall approach it, is the study of a cluster of interrelated concepts including those of function and goal. This section uses a discussion of the philosophy of functions as a springboard for the metaphysical investigation of goal-directedness that we begin in Section 3.

Talk of functions is pervasive. In everyday contexts we readily ascribe functions to the artefacts around us, such as a chair or toothbrush. Functions are also part of the explanatory practices of many branches of science, particularly in biology, medicine, and technology. Some argue that functions can even be ascribed in cases of physical, chemical, biochemical, and sociopolitical systems such the water and rock cycles, autocatalysis, Bénard cells, vitamin

B12, mineral species, and welfare systems.[9] If we are to take all this talk of functions at face value, we require an analysis of functions which explains how they arise in the natural world.

We start this section by offering a critical survey of two popular philosophical approaches to functions which are sometimes interpreted as trying to provide a *fully general* analysis of functions. These analyses received much attention in the 1970s and are still discussed today. We begin with the selectionist or 'etiological' approaches (e.g., Garson 2017, 2019a, 2019b; Mitchell 1993; Neander 1991; Wimsatt 1972; Wright 1973), before moving on to the causal role view (e.g., Cummins 1975). There is by now a vast literature on the philosophy of functions, with sophisticated versions of each approach. There are also hybrid theories which incorporate elements of both the selectionist and causal role views, such as the organizational theories of McLaughlin (2001) and Mossio et al. (2009). It is not our aim here to undertake a broad survey of all the various versions and their nuances;[10] we shall, however, try to go into enough detail to provide preliminary support for the idea that our theory of functions requires a sufficiently developed notion of goal-hood. In their basic forms, neither the selectionist nor the causal role analyses place much weight on the notion of goals. But this creates a problem on both sides for those who want their theory to provide a single, fully general account of functions across the board. On the one hand, selectionist analyses are arguably narrow because they apply mainly to the biological (and perhaps technological) domains, and it appears that in many domains function ascriptions are appropriate in cases lacking the appropriate causal history of selection (Section 2.2). This limitation motivates the causal role analysis, which accommodates functions that are not naturally selected. Unfortunately, however, the causal role theory arguably goes too far the other way and *over-generates* functions (Section 2.3).

In Section 2.4, we see how this dilemma can be avoided by employing the notion of goal-directedness. In order for some causal feature of an entity to count as a function, it must be one that contributes to the goal(s) of a system. We shall follow others in calling this the 'goal-contribution' theory of functions. Importantly, the concept of goal-directedness is not so narrow that it applies only to naturally selected functions and biological systems. On the contrary, as we shall see in Sections 3 and 4, influential analyses of goal-directedness (e.g., Babcock 2023) allow us to posit functions in wholly inorganic systems. What

---

[9] For a recent discussion of functions in biochemistry, see Bellazzi (forthcoming). On the alleged teleological features of mineral species, see Babcock 2023. For discussion of a classic debate on functional explanation in social science, see Schwartz 1993.

[10] More comprehensive critical appraisals of the various theories of function are provided in Nissen 1997 and Garson 2016.

the goal-contribution account provides is thus a general *metaphysical* analysis of functions which can be applied in many different domains of science.

To be clear from the outset, we certainly do not deny that the selectionist approach carves out an important type of function in biology (and, perhaps, technology) that delivers the kinds of 'backward-looking' explanations described in Section 1. In line with the explanatory desideratum (ED), any adequate metaphysical account of functions must be able to accommodate the backward-looking explanatory roles of selected functions. On the other hand, we would expect a fully general metaphysical analysis of function to be able to accommodate non-selected functions, if such there are. Again, we shall argue that the goal-contribution theory of function ticks all of these boxes. In Section 2.4, we discuss how the goal-contribution theory of function is able to accommodate selected functions in biology and their backward-looking explanatory roles. Section 2.6 also shows that a goal-contribution theory of function is compatible with a range of specific approaches to functions in the philosophy of biology. The key point is just that the goal-contribution account is a metaphysical theory that is broad enough to accommodate cases of functions that might fall outside the scope of the selectionist analysis. This means that the goal-contribution theory is not restricted to the biological domain and can accommodate talk of functions in many areas of science and everyday life. For that reason, it is hoped that the conclusions of this section will have broad appeal. Nonetheless, substantial difficulties arise in Section 3, where we attempt to provide a more detailed metaphysical analysis of goal-directedness.

2.2 The Selectionist Theory

Selectionist or 'etiological' theories of function rely on the idea that a function claim should explain the very presence of the function bearer, such as a trait of an organism. On this view, the main point of a function statement is to provide an answer to the question 'why is X there?' This type of question is prominent in biology, and in that context such explanations are made possible by the fact that instances of certain traits had effects in the past which explain their occurrences today. Such explanations thus take the form of a causal-historical story about natural selection (e.g., Bourrat 2021; Garson 2017, 2019a, 2019b; Millikan 1984; Neander 1991). This is the dominant and widely held theory of function within the philosophy of biology.

Without doubt, the selectionist approach carves out a very important class of functions. And a clear attraction of the selectionist theory is that it is thoroughly naturalistic. Rather than being grounded in, say, the intentions of a divine designer, teleology is grounded firmly in evolutionary biology. The theory is also able to ground the normative aspect of selected functions in a transparent way, in line with the normative desideratum (ND): the reason why, say, a heart should beat is that it is by virtue of their beating that hearts exist in the first place (see Garson 2019a: Ch. 7 for further discussion of this example). Thus, the selectionist theory also promises to shed light on how normative truths can be grounded by the natural world in at least some cases.

Abstracting away from the details of evolutionary natural selection, Wright proposed one of the first etiological definitions of function, which had the following two conditions:

'The function of X is Z means

a) X is there because it does Z,
b) Z is a consequence (or result) of X's being there' (1973: 161).

Interestingly, this definition is general enough that it might also apply in some non-biological cases. For instance, the function of an artefact like a chair might explain the chair's existence in the following sense: it is on the basis of certain characteristics (e.g., being comfortable to sit on) that chairs like this have been selected by agents like us for production. Indeed, artefactual functions look like teleological properties *par excellence* given that artefacts are designed with purposes in mind by agents like us. So, it appears to be an advantage of the selectionist analysis that it can subsume both the biological and artefactual cases. Nonetheless, it is arguable that the selectionist analysis still does not tell the full story about functions, because it applies only to functions which have just the right kind of causal history. For example, if there are (or could be) functions which are acquired accidentally rather than through a process of selection, then the selectionist approach will not apply to them. We shall explore possible examples in this Section.

Another noteworthy feature of selectionist functions is that they are somewhat contingent and extrinsic. Assuming that the causal history of the world is contingent, it is thereby contingent that a characteristic performs a certain function. This also means that, in such cases, function performance is not fully a here-and-now feature of an item. The attribution of a function to an item can only be made retrospectively once items of that kind have been reproduced on the basis of their effects in the past. Some have found this extrinsicality and contingency to be counterintuitive (e.g., Bigelow and Pargetter 1987: 187–188; Christensen and Bickhard 2002; Mossio et al. 2009: 921). Some have also found it objectionable that the selectionist theory allows us to say that a current trait has certain functions (based on ancestral natural selection) even though it no longer *performs* those functions for its possessors

(Bigelow and Pargetter 1987: 196).[11] Possible examples of such 'vestigial' features include the pelvis of the whale and the eyes of certain fish who live in darkness.

I do not think that, by themselves, these objections are decisive. The selectionists are obviously happy to accept the extrinsic, historical, and contingent nature of functions and will be inclined to dismiss intuitions to the contrary. And as noted already, the selectionist approach to functions is deeply entrenched in the philosophy of biology. Nonetheless, as metaphysicians we are entitled to ask whether *all* functions, across all scientific domains, have to arise through processes of selection. If the answer is 'no', then we must accept that the selectionist approach applies only to a proper subset of all possible functions, albeit a large and important subset.

Not only does it appear coherent to imagine functions that lack an appropriate causal history of selection, but arguably there are actual examples of functions that lack a selective history, or at least cases in which there is no evidence of natural selection. Godfrey-Smith (1993: Sect. 3.3) goes as far as to suggest that there is now some consensus on this matter. Anatomy and morphology (Amundson and Lauder 1994), biochemistry and neuroscience (Godfrey-Smith 1993), and even ecology (Bouchard 2013), have all been identified as areas in which the functions of primary concern are often *not* the result of natural selection.[12] Consequently, many who are otherwise sympathetic towards the selectionist theory have softened their position and accept that the theory might not cover all possible cases of function attribution in all domains of science.[13]

Boorse (1976, 2002) helpfully collates a range of actual and possible counterexamples to the Wright-type etiological analysis, some of which come from within biology itself. Among these are cases of so-called 'exaptation' proposed by Gould and Vrba (1982). Gould and Vrba introduced the notion of exaptation to modern biology after noticing that many features of organisms are co-opted by those organisms and their descendants for uses for which those features were not naturally selected. One of Boorse's favourite examples of possible exaptation is of sea turtles who dig egg holes with their flippers: surely this is a current function of the flippers, regardless of whether the flipper was originally selected for that function (Boorse 2002: 66). Gould and Vrba (1982: 7–11) present other

---

[11] Bigelow and Pargetter (1987: 190) also argue that if we accept a selectionist analysis of functions, then backward-looking functional explanations become somewhat trivial. This is because, by definition, functional characteristics are those which have persisted by virtue of their previous effects. However, see Mitchell (1993: 253–254) for a response.

[12] See also the recent debates about ENCODE in genetics, for example, Doolittle 2013, Germain et al. 2014.

[13] For further discussion of this issue, see Brandon 2013, Garson 2018, Griffiths 1993, Melander 1997, Millikan 1989, Neander 1991, Preston 1998, Shaffner 1993, and Walsh and Ariew 1996.

possible (and surprising) examples of exaptation. For example, there is a conjecture that the enlarged feathers on the hands of birds were originally selected for insect-catching rather than flying. But even if this were so, surely we should not revise our judgement that the contribution to flying is a function of wing feathers. Another theory has it that in vertebrates, bones were originally selected for their ability to store minerals such as phosphate, rather than their ability to support and protect.[14] And note that even if the historical details in these cases turn out to be incorrect, the *mere possibility* of unselected functions is sufficient to cast doubt on the selectionist theory as a fully general analysis.

To be clear once again, the proposal is not that function statements *never* play the kind of important explanatory role that selectionist theories focus on. Backward-looking functional explanations are clearly crucial in some areas of science – particularly in evolutionary biology. The point is just that *not all* function statements have to play this role. What this ultimately suggests is that function statements can be used to answer more than one kind of 'why'-question. According to Garson (2018), there are at least two different kinds of 'why'-question in biology and related disciplines such as neuroscience and psychology (see also Mitchell 1993: 258–259 and Tahko 2020: Sect. 2). Understanding this distinction provides the key to understanding the different explanatory roles of function statements and is therefore crucial for satisfying the explanatory desideratum (ED). The first type of question is the 'why-is-it-there?' question (Garson 2018: 1102), and this is where selected functions are called for, for reasons discussed already in this section. However, another why-question is what Garson calls the 'what-does-it-do?' question, which we ask when we want to know how a trait or mechanism contributes to a certain system-level capacity. In order to answer questions of this latter kind, what we require are function claims that focus more on the causal role(s) of the function bearer. Even within a *particular* branch of science, it seems that both kinds of questions may come into play (Forber 2020: 277). With this in mind, we shall now explore another notion of function that focuses more on the causal role(s) of its bearer.

2.3 The Causal Role Theory

According to the causal role approach, what makes talk of functions appropriate in both selected and unselected cases is that the functions play certain *causal roles*. Indeed, the second condition of Wright's analysis (Section 2.2) acknowledges this,

---

[14] Other possible examples: When discussing the work of Colin Pittendrigh, Cartwright (1986: 208) notes how, arguably, the biological clock and its associated functions piggybacked on circadian oscillations that already existed. James Miller also tells me there is a Chomskyan hypothesis that the syntactic elements of speech were not directly selected but rather were a by-product of earlier evolutionary changes in cognition.

since it concerns the contribution made by a function bearer. The causal role approach puts this idea at centre stage. According to this theory, being able to play an appropriate causal role is the necessary and sufficient feature of a function, while being naturally selected is a feature that some but not all functions have.

According to Cummins (1975), by focusing too much on the need to explain the *presence* of a function bearer, selectionist theories overlook a core feature of functions across the sciences and even within biology itself. To use one of Cummins's illustrations, 'The function of the contractile vacuole in protozoans is elimination of excess water from the organism' (1975: 749). In the example, the presence of the contractile vacuole allows there to be a semi-permeable membrane around the organism, which is necessary for the movement of waste out of it. According to Cummins, the explanatory value of this function ascription is that it informs us about what is responsible (in part) for the elimination of the relevant organism's excess water, and in turn explains how the presence of the contractile vacuole contributes to the survival of that organism. Importantly, however, this is *not* to say that survival value is constituted by natural selection. For Cummins, the traits of organisms are determined by their genetic plans, as are their survival-enhancing roles. What natural selection does is merely react to these genetic plans, 'weeding out' the bad ones (Cummins 1975: 751). Whether or not a genetic trait is subsequently weeded out depends on whether it has survival-enhancing capacities *in the first place*. Thus, Cummins's theory shifts the focus of function statements towards the *dispositions* of a trait.[15]

Cummins's work has undoubtedly provided helpful insights, particularly in emphasizing the causal contributions that functions make. However, one might still have doubts about whether Cummins's causal role theory can itself deliver the whole story about functions. The problem, as Cummins acknowledges, is that just about any effect or activity contributes to the persistence of some condition of a containing system. For example, 'Heart activity ... keeps the circulatory system from being entirely quiet' (1975: 752): yet this is surely not a function of heart activity.[16] So, prima facie, Cummins's theory faces a problem of overbreadth: if not all dispositions are functions, then a purely dispositional account inevitably leaves us with too many functions. And this prima facie impression is reinforced

---

[15] Whether such dispositions are always *intrinsic* is a tricky question. As a referee has helpfully pointed out, some contemporary theories of genetics discourage the idea that genetic properties are intrinsic (see, e.g., Griffiths and Stotz 2013 and Bellazzi 2022). More generally, it is far from clear that all genuine dispositions are intrinsic, even though many clearly are (McKitrick 2003, 2018).

[16] I note that Mumford might disagree, since he maintains that all dispositions are indeed functions (1998: 200–202). However, Mumford is engaging in a rather different project. We are concerned with function concepts in the sciences, whereas Mumford develops a philosophical notion of function that can be used to explicate the theoretical role of disposition ascriptions (1998: 198). This latter sense of function is fine as a philosophical term of art, but it should not be conflated with the more scientific notion of function that we are examining.

when we look at different areas of science, since it does indeed seem fair to say that not all dispositions are treated as functions. As Bigelow and Pargetter note, a meteorologist would not say that a function of mist is to produce rainbows (1987: 184), nor would a geologist think that a function of rocks in a river is to widen a river delta (Kitcher 1993: 390), even though mists and rocks do have the corresponding dispositions. To use another common example, it also does not seem right to say that oncogenes have the function to cause damage to organisms, even though they do have such dispositions.

An obvious response for the causal role theorists is to reply that a genuine function is one that contributes to some *other* function that its containing system has as a whole. We might then emphasize that oncogenes thwart rather than aid these overall functions. However, this merely pushes the problem back a step, for we have appealed to a further function concept, and if that further function is then clarified with reference to yet another function or a further containing system, we face the threat of a vicious regress, as Cummins acknowledges. For example, we might try to explain the function of a heart with reference to the overall function of the circulatory system; but if the function of the circulatory system in turn depends on further functions of an organism, then '[e]ither we are launched on a regress, or the analysis breaks down at some level for lack of functions, or perhaps for lack of a plausible candidate for containing systems' (1975: 752).

Cummins's response to this problem is somewhat surprising. In order to draw the distinction between functions and mere accidental effects without appealing to further functions, Cummins appeals to the 'analytical context' in which the functional ascription is being made: 'It is appropriate to say that the heart functions as a pump against the background of an analysis of the circulatory system's capacity to transport food, oxygen, wastes, and so on, which appeals to the fact that the heart is capable of pumping' (1975: 762). However, relative to some other capacity of a containing system, the functions of that system's components may well be different. On this view, what counts as a function (as opposed to a mere disposition) ultimately depends on the theoretical interests of those ascribing the function. The aforementioned regress is terminated because, relative to a certain analytical context, it is simply taken as a given that a system has a certain capacity. And what is taken as given depends on the interests of the observer. If the context shifts, then some other capacity can come into focus and what counts as a contributing function may then change. This openly epistemic, contextualist view has been influential in recent causal approaches to function, such as the neo-mechanist account of Craver (2013).[17]

---

[17] For another contextualist take on functions, see Prior 1985. See also Tahko (2020), who in a discussion of proteins suggests that their functional classification brings a risk of interest relativeness.

Unfortunately, this epistemic and contextualist approach to functions leads to the obvious worry that without the perspectives or interests of observers, things *do not really have functions*, even though they have relevant dispositions. Some might welcome this conclusion, especially those who are generally sceptical of metaphysical teleology. However, others have found this alleged interest-relativity of functions implausible. For example, Bigelow and Pargetter urge that 'biological structures *would* have had the functions they do have even if we had not been here to take an interest in them at all' (1987: 183), and also that 'some of the effects of structures that we take an interest in have nothing to do with their function' (1987: 184).

Notice also that this epistemic approach is likely to weaken the normative credentials of functional explanations. As highlighted by our normative desideratum (ND) in Section 1, a theory of function should accommodate the normativity of functions, including the distinction between normal function and accidental function. However, insofar as Cummins's theory retains the normative dimension of function statements, this too will be contextual. Relative to our interest in capacity c of a system, a certain component might have a certain normal function, with other behaviour being regarded as merely accidental. But relative to our interest in capacity c' of that same system, it may be that the normal function of that same component is different. On this account, it is difficult to see how the natural world *in itself* fully grounds these normative claims.

For those seeking a more robust basis for function ascriptions in science, the selectionist approach has a significant advantage over the Cummins-type causal role theory, because the former can at least provide an objective, observer-independent basis for normative claims about many functions in terms of what an entity has been naturally selected for (see e.g., Bellazzi forthcoming: Sect. 5.1, Garson 2017, 2019a; Millikan 1989; Neander 2017). The selectionist approach therefore fares well with the normative desideratum (ND). Nonetheless, the causal role theory does contain some important insights. In the following subsection we shall explore a theory that preserves some of the insights of Cummins's theory while offering a more robust theoretical foundation for function claims and their normative implications. The theory in question is the so-called 'goal-contribution' approach to functions. In the course of our discussion, we shall also explain how a goal-contribution approach can accommodate the sorts of 'backward-looking' functional explanations that philosophers of biology are particularly interested in.

2.4 The Goal-Contribution Approach

In the 1970s, Boorse (1976), Nagel (1979: Ch. 12), and Adams (1979) rejuvenated a view known as the goal-contribution theory of functions, which has its roots in earlier works in philosophy of science by, among others, Beckner (1959), Braithwaite (1953), Sommerhoff (1950), and Nagel (1961). More recently, the goal-contribution account of function and teleology, or variants thereof, have been endorsed to varying degrees by, among others, Babcock (2023), Boorse (2002), Lee and McShea (2020), McShea (2012), Schaffner (1993), and Trestman (2012).

The goal-contribution account is sometimes regarded as an extension of the causal role approach, insofar as it focuses in part on the causal contributions that a functional item makes to a system. Indeed, Nagel's goal-contribution analysis contains an explicit causal component according to which a functional entity makes a necessary contribution to a certain effect. However, unlike Cummins's account, the goal-contribution theory insists that an entity can only be said to perform a function if it contributes specifically to a *goal* of a system, understood as a characteristic state that the system strives to bring about. Thus, the goal-contribution account invokes the core concepts of *systems* and their *goals*, about which we shall say more in a moment and in Section 3. The important point for current purposes is that the goal requirement promises to impose an objective restriction on the kinds of causal contribution that can count as functional. The idea is that this restriction will allow us to avoid Cummins's overbreadth problem discussed earlier, and thereby help to satisfy the desideratum of extensional adequacy (EAD) introduced in Section 1. As Adams puts it:

> So, e.g. we can avoid attributing to the moon the function of making the tide come in and go out. Although this surely is an effect of the moon, it simply does not warrant an ascription of function. We must have a way to block such ascriptions, and wedding the ascription of functions to goal-directed systems provides the means to do it. (1979: 496)

Referring back to the truthmaking desideratum (TD) of Section 1, the proposal is that certain systems have objective goals, and the causal contributions made to those goals by the relevant function bearers are what make the functional truths true. Given that Boorse has offered one of the most detailed and general goal-contribution analyses of function, I shall focus mainly on his theory and see how the theory applies in different cases of function. In Section 3, we shall then move on to Nagel's specific 'cybernetic' understanding of goals.

For Boorse there is more than one kind of function ascription, and the one that is most widely applicable has the form 'X is performing function Z'. This kind of function statement is widely applicable, in part, because it is sometimes applied even if the function in question is performed only once in a somewhat accidental way. For example, one might say of a particular pocketbook that it performed the function of stopping a bullet hitting a soldier, even if this only happens once (Boorse 1976: 80). Alternatively, one might say that a rock performed the function of a coffee table for a person on a particular occasion. A function claim like this cannot be generalized but rather applies at a particular point in time. Hence, Boorse's analysis of function performance involves time variables (t). And due to the important requirement of systemic goal contribution, the analysis also includes variables relating to systems (S) and the systemic goal (G) to which a functional item (X) contributes. In the cases above, the goals are those of the people involved. This leads to the following analysis:

> 'X is performing the function Z in the G-ing of S at t, *means*
> At t, X is Z-ing and the Z-ing of X is making a causal contribution to the goal G of the goal-directed system S.' (Boorse 1976: 80)

Since this definition employs the important concept of a system, we should say a little more about what that means. The term 'system' is broad and can be applied to any group of things that are connected in some way. Talk of systems is pervasive in many branches of natural science, which are typically concerned with causal or nomic connections between concrete, spatiotemporally located things. At school we learn, for example, about planetary systems, weather systems, and immune systems. However, the term is not restricted to the concrete realm. We also find talk of systems in the rational sciences: for instance, there are different systems of logic, where propositions are connected via various rules of inference. Given that the concept of a system is so broad, it is not easy to provide a general definition.

In the current context, however, we can at least say that the systems of interest are concrete ones with causally interacting parts. One of the important lessons of Cummins's work, I take it, is that the concept of function is, first and foremost, a causal one. That is not to say that all concrete systems will count as being goal-directed, however. For example, it is far from clear what the goal of, say, the solar system could be. On the other hand, we need not restrict the notion of goal-directedness only to organic systems. Although the notion of a goal is often associated with the concept of intention or desire, goal-contribution theorists typically do not want to restrict talk of functions and goals only to systems capable of having intentions. We discuss this point further in Section 3. Boorse also urges that it would be a mistake to align the concept of a goal too closely with either reproductive fitness, usefulness, or the good, as certain philosophers have sometimes tried to do (1976: 77). The flexibility that Boorse endorses is, I think, to be welcomed, and I shall say more about it in Section 2.6. The relevant point for current purposes is that philosophers could agree about the goal-contribution analysis of functions but have further in-house disagreements about which cases satisfy the schema.

After defining the notion of function performance, Boorse introduces a more general kind of function statement. Intuitively, not all functions performed by

X are its *characteristic* functions. Although someone's pocketbook might perform a useful bullet-stopping function on one occasion, this is not *the* general function of that book. Rather, the general characteristic function of a book is, I assume, to be read. Other general function locutions include '*X has the function Z*' and '*A function of X is Z*' (Boorse 2002: 71). How, then, should we understand the difference between the more general function locutions and the specific statements which are merely to do with a function performed – perhaps somewhat accidentally – by X on a particular occasion? The thrust of Boorse's answer is that the appropriateness of a general function statement is largely to do with how often the relevant function is performed by X (1976: 80–81): 'If Schaffner's mutant bursa of Fabricius (1993: 388) blocked viral infections throughout its owner's life, it would be quite natural to call antiviral defense the bursa's function in this man' (2002: 71). Returning to the earlier example, blocking bullets not *the* general function of a particular book because it is not something the book will do for people on a consistent basis.

Notice also that the goal-contribution approach allows that something can simultaneously contribute to the goals of more than one system. So, when we speak of *the* general function of a particular item, what we really mean is *the function of the thing within such-and-such system*. For instance, a bird's capacity for eating insects is clearly a function, but when we ask which goal the insect-eating is serving there are several possible answers depending on which biological system we are considering: eating insects not only contributes to the survival of the individual bird in question, but also arguably to the survival of its species, of its genes, or the equilibrium of the insect population (Boorse 1976: 83). However, this does *not* mean that the goal-contribution theory leads to a merely perspectival or human-dependent account of functions. To secure the objectivity of function, what is important is that judgements about the goals of a system are grounded by the nature of the relevant system itself. There is more to be said about this issue and we shall return to it in Sections 3 and 4, where we examine the nature of goals and goal-directedness in more detail.

Let us now consider some of the advantages of a goal-contribution analysis of function, of which there are several. First, the goal-contribution analysis of function can accommodate the selected functions discussed in Section 2.2. In goal-theoretic terms, naturally selected biological functions are ones which previously contributed to the goal of the survival of the genes that generated them. Importantly, the theory can also accommodate the etiological explanations associated with such functions, thereby satisfying the explanatory desideratum (ED). The disposition of a trait to contribute to the goal of an organism (i.e., fitness) will, in conjunction with evolutionary theory and environmental facts regarding habitat, explain the prevalence of present tokens of that trait

(Boorse 2002: 79). In other words, given that *past* tokens of the trait contributed to the goal of fitness, such tokens will tend to propagate. The main difference between this account and the selectionist theory is just that the goal-contribution approach does not write such explanations into the very *analysis* of function ascriptions. And given that not all functions are selected, this looks like a benefit.

Importantly, although the goal-contribution theory has considerable flexibility, it is still less flexible than Cummins's version of the causal role theory. We saw in the previous section that Cummins's theory faces the problem of overbreadth. However, it looks as though the goal-contribution theory has the resources to weed out the common counterexamples facing the causal role theory. In general terms, the counterexamples will be avoided if the items in question are not a part of the relevant system or if the effects in question do not really make a causal contribution to a genuine goal of a system. A glance at some of the counterexamples mentioned earlier suggests that the goal-contribution restriction does indeed help. Since oncogenes cause damage to organic systems, they cannot be said to contribute to the system's goals – on the assumption, at least, that the biological goal of life is health and fitness (Boorse 2002: 73, 76). And although science regards rocks as being parts of some systems, such as the rock cycle, river water systems are arguably not among them, implying that it is not really a function of rocks in a river to widen a river delta. Of course, questions remain about what it takes to be a goal-directed system and we shall address these in Sections 3 and 4. I also leave it as an open question as to whether the goal-contribution account of function is itself counterexample-free.[18] The important point for current purposes is just that the goal-contribution account of functions is more discerning than the Cummins-type causal role theory, which looks like progress.

2.5 Goal-Contribution Functions and Normativity

In this section, we shall briefly look at some of the details that Boorse (2002) adds to his theory in order to account for the normativity of functions. As we saw in Section 1, it is a desideratum of a theory of functions that it accommodates the normative force of various function statements (ND). The literature on selected

---

[18] A referee has queried whether a goal-contribution theorist can accommodate the possibility of a goal-directed system altering its goals or functions due to environmental changes. In Section 4, we shall develop a theory of goal-directedness using a realist view about dispositions, and within that framework this question boils down to whether some genuine dispositional properties are partly extrinsic. I see no obvious reason why there cannot be such dispositions, but I do not have the space to discuss the issue in detail. Again, see McKitrick 2003 and 2018 for further discussion of extrinsic dispositions.

functions fares well in this respect: the normal or proper function of an item is that for which it was naturally selected. However, if not all functions are selected, we need some other general way of accommodating the normativity of functions. In order to do so, recall, the analysis must at least be able to draw a distinction between non-function and malfunction, and a distinction between normal (or proper) function and mere accidents (Wright 1973: 165; see more recently Garson 2019a: Chs. 7 and 8).

Perceptive readers may have noticed in the previous subsection that Boorse's more general kind of function statement already generates a function–accident distinction. X's contribution to some system's goal can happen only once and be fortuitous. In that case, X has performed a certain function just on that particular occasion. Although such cases of function performance exist, it does not follow from this that X generally *has* that function. A function performed is not the same thing as a function possessed (Boorse 2002: 71). However, if X makes the same kind of goal contribution often enough, we would be inclined to regard the contribution as being a normal function of X. This general notion of function applies to a particular *token* item. But in science, we also make important generalizations about what is typical of a *type* or species of thing. Often, when we say that someone's kidney is functioning normally, we mean not only that this particular kidney has the function in question, but also that this function is typical of this *type* of organ. In order to accommodate attributions of normal or proper function in such cases, we must appeal to type-level concepts.

In other work, Boorse analysed normal functions in medicine using the type–token distinction and facts about statistically typical behaviour (e.g., 1977: 554–563). On the assumption that the ultimate biological goal of an organism is fitness, we may say that a normal function of a given type 'is analysable as an output within a statistically typical range of contributions to survival and reproduction by tokens of that type in an age group of a sex of a species' (Boorse 2002: 72). Thus, at the level of types, we may say that some causal contribution to a systemic goal is a normal function if this function is typical among the members of the relevant type. Such normative judgements require us to look beyond the behaviour of an individual item X and examine the relevant statistical distributions of behaviour concerning other tokens of X's kind.

To be clear, the fact that a token item X is *performing* a function is determined *locally*. However, if the function performed is not species-typical, then we may conclude that the function is not normal or proper for that type of entity. Boorse concedes that he cannot specify precisely how many times a function needs to be performed within a species or kind in order to make a function normal. However, Boorse notes that this sort of imprecision is a feature of other theories of functions too. For example, if we accept a general selectionist analysis of function, we are left with the similar question of how often a contribution of a trait must make its bearers better able to reproduce, in order to be regarded as a normal or proper function.[19]

There is, however, a fairly obvious problem case for this statistical account of type-level normal function. This is what we may call, following Neander (1983: 80–81), the problem of species-typical fortuitous consequences. Well-known examples of widespread fortuitous effects include the common ways in which noses support eyeglasses and the common ways in which hearts produce sounds that help doctors to diagnose various illnesses. Both of these effects can contribute to the well-being of an individual, and well-being is surely one of the goals of an organism. However, if we arrive at a point at which most people wear glasses, it seems Boorse will have to say that supporting eyeglasses has become a normal function of the nose. Neander takes this to be the wrong verdict.

Boorse's response (2002: 88) is to bite the bullet and accept that, in these cases, it would indeed be acceptable to say that the human nose has taken on a new function. Boorse cites other examples that are analogous, including the sea turtle example mentioned earlier. At present, sea turtles often use their flippers for egg-hole digging, thereby aiding reproduction. It is not plausible to think that these flippers were naturally selected for this purpose, yet it seems far from absurd to say that egg-hole digging has become a normal function of the sea turtles' flippers. If the goal-contribution theory is attractive in other independent respects, then perhaps Boorse is entitled to stand his ground on this issue.[20]

Regarding the non-function/malfunction distinction, the challenge is to make sense of how an item can be deemed defective. For example, neither a severely diseased heart nor a wooden spoon has the ability to beat and pump blood effectively. Nonetheless, we are inclined to say there is an important difference between these two cases. The diseased heart is defective in the sense that it should beat but doesn't: it is malfunctioning. How, then, can a goal-contribution theorist account for such malfunction? Boorse's answer (2002: 88–89) is to draw once again on the type–token distinction and the notion of normal functioning. Neither the severely diseased heart nor the wooden spoon can *perform* the function of beating. Nonetheless, this particular heart is a member of an anatomical type – *being a heart* – whose normal function is to beat. This is to

---

[19] Moreover, Boorse's statistical account of species-typicality is more broadly applicable, since it applies not only to types with a certain evolutionary history.

[20] A referee points out that in Section 2.3, we complained that Cummins's causal role theory over-generates functions, and so one wonders whether Boorse is in the same boat. In response, I would emphasize that Boorse's goal requirement places an important restriction on what can count as a function. For example, goal-contribution theories can deny that oncogenes have the function to cause damage to organisms, because such damage contravenes the biological goal of survival.

say that, statistically speaking, tokens of this type typically beat for the benefit of some goal. In contrast, a wooden spoon is not a member of a type that typically beats. Hence, we may rightly draw the verdict that in terms of beating, the diseased heart is defective while the wooden spoon is not.

This is not quite the end of the matter, however. We have just seen that Boorse's analysis of normal function faces the problem of species-typical fortuitous consequences. The account of malfunction faces an analogous problem. It is, for example, conceivable that there could be a devastating pandemic which results in, say, very few hearts beating. In such a scenario, it would not be species-typical for those hearts to beat, and yet surely it would remain the case that the non-beating hearts are defective or dysfunctional. We would still believe that the function of a heart is to beat even though most hearts would no longer happen to perform that function. How, then, can goal-contribution theorists like Boorse accommodate this fact? Melander (1997), Millikan (1993), Neander (1991, 2002), and Plantinga (1993) have all raised this kind of worry using various examples, some actual and some merely possible.

Boorse's response to such counterexamples is lengthy (2002: 92–103), and so I shall focus on what I take to be the main thrust of his solution. Boorse replies that we should be flexible about what counts as the relevant reference class when assessing statistical normality. In particular, it is in many cases sensible to examine an extended time-slice of the kind or species. Boorse thinks this is an inevitable feature of the approach, because 'If the whole earth went dark for two days and most human beings could not see anything, it would be absurd to say that vision ceased to be a normal function of the human eye' (2002: 99). Hence, even if the devastating pandemic occurred, and most current hearts failed to beat, we could still insist that the function of the human heart is to beat given that this is what hearts have typically done in the extended past. Hence, if Boorse's account utilizes the relevant historical data, it can accommodate the idea that most hearts are malfunctioning in the pandemic example (for further discussion see Garson and Piccinini 2014 and Garson 2019b).

This does, however, leave us with the question of just how far we are meant to go back in history in order to fix judgements about malfunctioning. In epidemiological contexts, it is far from clear how far back our reference classes should go (Smart 2016: Ch. 2). In Boorse's work, the answer appears to be anywhere between two lifetimes and a millennium (Giroux 2015: 185). Again, this brings to light the fact that normalcy and typicality are not precise concepts. I shall leave readers to judge how problematic (if at all) this imprecision is. But in doing so, we should not lose sight of the many benefits of the goal-contribution theory. There is, I think, no guarantee in philosophy and science that all our theoretical concepts can be defined in a perfectly precise way.

In the final subsection of this section, we shall outline a further attraction of the goal-contribution theory, which is that it is compatible with many of the more fine-grained theories of function in the recent literature on philosophy of biology.

2.6 Compatibility of the Goal-Contribution Approach with Recent Theories of Function

A benefit of the goal-contribution account of function, touched upon already, is its generality and hence flexibility. It allows that different kinds of system, including non-biological ones, may have very different goals, and it remains neutral about the specific details in each case. Indeed, in Section 3 we shall discuss cases of goal-directedness involving cybernetic systems. We have also seen how the goal-contribution theory can accommodate the sorts of backward-looking functional explanations that are prominent in evolutionary biology. In the current subsection, we shall see that the goal-contribution theory is also compatible with various other popular theories in recent philosophy of biology regarding the nature of functions. In what follows I give three examples, namely, the survival and reproduction account, the organizational theory, and the neo-Aristotelian approach. To this we could also add the theological approach to teleology that we mentioned briefly in Section 1, since the goal-contribution theory is compatible with the further idea that the goals of, say, organic systems like us are ultimately determined by some divine power.

According to the survival and reproduction account, biological functions are those which contribute, or are disposed to contribute, to biological fitness in various ways (e.g., Bigelow and Pargetter 1987; Canfield 1964; Ruse 1971). It is easy to see that this approach could be interpreted in goal-theoretic terms, because this is precisely Boorse's view in the case of biology and medicine (2002: 64, 108). In Boorse's view, the survival and/or reproduction of an organism are precisely its 'apical' goals (2002: 76), and each of its functioning subsystems contributes to these ultimate goals in one way or another. Elsewhere, the goal-directedness of developmental biology has also been emphasized by some philosophers of science. For example, Austin highlights the goal-directed nature of models in dynamic systems theory, which privilege certain end-state morphologies (Austin 2017: 200).

Organizational or 'organismic' accounts of biological function have also become prominent in recent literature. Such accounts come in various forms which overlap in various ways with the causal role and survival and reproduction accounts. However, the idea is roughly that an item has a biological function if it plays an appropriate role in the maintenance of the system on which it reciprocally depends (e.g., Barandiaran and Moreno 2008; Kertész and Kodaj 2023; McLaughlin 2001; Mossio et al. 2009; Mossio and Bich 2017). This reciprocity requires a certain kind of organizational complexity, which all self-maintaining biological systems share. Mossio et al. (2009) spell out this complexity in terms of the notions of 'organizational closure' and 'organizational differentiation' (2009: 824, 826). A system is organizationally closed when there is a 'circular causal relation between some macroscopic (or higher-level) pattern or structure and the microscopic (or lower-level) dynamics and reactions' (2009: 824). This idea is closely related to the biological notion of autopoiesis developed by Maturana and Varela (1980), whereby biological functions are said to be continually realized and regenerated via the organization of organisms (see also Meincke 2019). The important point for current purposes is that organizational closure can be spelt out naturally in goal-theoretic terms. Mossio et al. (2009: 824) indicate as much when they say that 'in an organizationally closed system the goal states are the stability points (or set of points) through which the system can exist'.

Austin and Marmodoro's neo-Aristotelian account of organisms (2017) also aims to do justice to these organizational features (see also Oderberg 2007: Ch. 8). The novelty of their account is that such organization is underpinned by neo-Aristotelian realism about powers, which we discuss further in Section 4 when considering the metaphysical foundations of goal-directedness. When applying the powers metaphysics to the specific case of biological functional, Austin and Marmodoro propose that organisms have self-directed 'structural powers' which engage in 'cyclopoietic' activity: 'in traversing a kind of causal loop among the constituents of an organism, each is tied together in a continual diachronic cycle of co-production and maintenance' (2017: 171). Austin and Marmodoro go on to spell out the nature of these structural powers in goal-theoretic terms: 'The self-directedness of structural powers therefore consists in the *recursive* nature of this unifying activity—they are "self-orientated" precisely because the goal of that activity is to establish the cyclical perpetuation of its own operation' (Austin and Marmodoro 2017: 171–2). More recently, Paolini Paoletti (2021a) has developed a more general powers-based account of function, and he explicitly acknowledges that the theory is compatible with a goal-contribution interpretation of functions (2021a: 128). Again, we shall return to the important topic of powers in Section 4.

Finally, there is another sense in which the goal-contribution account is flexible—a sense which brings us to key questions to be addressed in Sections 3 and 4. So far, we have said little about whether goal attributions commit us to the existence of irreducible teleological properties in the world. As we shall see in Section 3, Nagel's goal-contribution theory of functions (1979: Ch. 12) involves

Teleology 25

an analysis of goal-directed systems that has its roots in the reductive cybernetic tradition. Although Nagel thinks that our concept of goal-directedness captures an important class of complex systems, his view is that such systems can ultimately be characterized in non-teleological, mechanistic terms. So, we may regard Nagel as a deflationist or reductionist about teleology: as we put it in Section 1, Nagel thinks the world is teleological with a small 't' rather than a big 'T'.

However, we shall see that matters are not straightforward for the Nagelian reductionists. Objections raised by Scheffler (1959, 1963); Ehring (1984a); Nissen (1981, 1997) and others show that it is difficult for cybernetic theorists to explain goal-directed behaviour in cases where the relevant goals are not achieved. In Section 4, we argue that a realist theory of powers offers resources for overcoming this problem. According to that proposal, goal-directedness can be understood as a type of end-directedness that is grounded in certain powers of a complex, 'directively organized' system. Importantly, it is arguable that powers are themselves irreducibly teleological. If that is correct, then the prospects for eliminating teleology from our metaphysics are dim. Despite the array of philosophical literature suggesting the contrary, the world may well be teleological with a big 'T'.

3 Goals

3.1 A Reductive Approach

As noted earlier, Boorse's goal-contribution analysis of functions has its roots in earlier work by Nagel (1961, 1979: Ch. 12) and others in the so-called 'cybernetic' tradition. According to Boorse and Nagel, function ascriptions are made relative to goals. As Nagel puts it: 'A functional statement of the form: a function of item *i* in system *S* and environment *E* is *F*, presupposes ... that *S* is goal-directed to *some* goal *G*, to the realization or maintenance of which *F* contributes' (1979: 312). If we accept this view, then we cannot fully understand the concept of function unless we understand the nature of goals and goal-directedness. Thus, in this section, we focus more directly on the concept of goal-directedness itself. In metaphysical terms, we may ask: What are the truthmakers for ascriptions of goals and goal-directed behaviour to a system? As far as I know, Boorse himself does not say a great deal about this important issue, perhaps because he thinks the earlier cybernetic literature has already shed light on the matter, notably via Nagel's 'system-property' theory of goal-directedness. The cybernetic approaches offer an empirical, reductive, and largely behaviouristic account of goal-directedness: the truthmakers for truths about goals and goal-directedness are just patterns of behaviour of a certain sort, which are generated by mechanisms of a certain sort.

Nagel's work has an openly reductive flavour, in the sense that talk of teleological goals is reduced to mechanistic language which does *not itself* employ teleological terms. This, in turn, has a knock-on effect regarding the goal-contribution account of functions, because functions can then be understood in terms of their contributions to systemic effects (the 'goals') which can, in principle, be characterized in entirely non-teleological terms. We shall explore the details in what follows. The upshot is that one cannot help thinking that if Nagel's theory is right, then although talk of goals or functions might be convenient, and of heuristic value in biology and other special sciences, we should not think that the world is *inherently* teleological. Teleological explanations thus become neither indispensable nor fundamental. For Nagel, the difference between teleological explanations and their non-teleological equivalents is ultimately one of emphasis (1961: 422). However, as we shall see, the cybernetic analysis of goal-directedness faces several counterexamples and problem cases, the most serious of which is *the problem of goal failure*. In seeking to overcome this problem, it is far from clear that we can avoid positing irreducibly teleological properties after all.

Let us start with an important clarification that was implicit in our discussion of the goal-contribution theory of function. Although functions arguably depend on goals, goals and functions are not the same thing. Nagel (1961: 277) remarks that biologists sometimes use these terms interchangeably, but that it would be a mistake to do so across the board. In the sense that we have defined functions, something performs a function when it *contributes* to a goal of a system. But that is not to say that the item with the function is itself a system with a goal. For example, very simple artefacts like coat pegs or fridge magnets perform functions when utilized by us in our goal-directed behaviour, but they are not by themselves complex enough to exhibit goal-directed behaviour. Moreover, at the level of a system as a whole, the system's goal may not be a function for anything because it may not contribute to a goal of some further system. Nonetheless, in the case of complex organic subsystems, their goal states *do* typically make contributions to the goals of a larger system, in which case the subsystem's goal states are themselves performing a function relative to some larger system. In complex organisms like us, there is a hierarchy of systems and subsystems at different levels of complexity, from genes through to cells, tissues, organs, and multiple-organ systems. Arguably, many of these subsystems have functions and goals which contribute to the functions and goals of a further system next up in the hierarchy. At the apex of this hierarchy is the organic system as a whole, and there appears to be some consensus in biology that the overall goal of an organism is fitness, that is, survival and perhaps reproduction (Boorse 2002: 70). Each subsystem can then be seen as contributing in some way to this overall goal.

So far, we have focused on biological examples, and in such cases it is quite natural to ascribe goals and speak of goal-directed behaviour – especially in relation to organisms capable of intentional thought. However, early goal-directed accounts of teleology were in fact inspired by technological developments of cybernetic machines, whose behaviour seems apt to be described as goal-directed (Rosenblueth et al. 1943). The much-discussed examples include homing missiles, thermostats, a steam engine's Watt governor, and, more recently, self-driving cars. The apparent goals in these examples are quite different: the homing missile's goal is to destroy some external target, while the goal of a thermostat is to maintain a certain temperature in response to external conditions. In each case, the behaviour of the system is *directively organized* towards a certain end. That is to say, both homing missiles and thermostats adapt their behaviour in varying conditions so that the goal-states can still be achieved. If a homing missile's target moves, the missile moves accordingly. If the temperature of a room rises above a certain level, the thermostat's feedback mechanism ensures that the heating system stops producing more heat.

One of the aims of the cybernetic programme was to unpack precisely what it is that cybernetic machines and organic systems have in common, and thereby explain and justify talk of goal-directedness in both cases. According to the cybernetic analysis, goal-directedness is largely an *intrinsic* feature of systems, grounded in their behavioural profiles. If successful, then, the programme would allow us to ascribe goal-directedness to systems independently of human aims and purposes.[21] And if truths about goal-directedness can, in this way, be grounded in mechanistic patterns of behaviour, we are left with a thoroughly naturalistic and reductive theory of teleology. If this approach is the correct one, then we do not need to account for goal-directedness in terms of mentalistic teleological concepts such as intentionality. Indeed, in order to signify their move away from irreducible teleology, cybernetic theorists often refer to goal-directed systems as being 'teleonomic' rather than 'teleological' (e.g., Pittendrigh 1958; Mayr 1988).

Given these naturalistic and reductive aims, the early cybernetic views are often regarded as taking a behaviouristic approach (e.g., Nissen 1997: Ch. 1), meaning that the teleological truths are grounded primarily in overt patterns of behaviour. It is in these terms that Rosenblueth et al. (1943) characterize their project, describing it as 'the behaviouristic study of natural events' (1943: 18).

---

[21] This is not to deny that the *activation* of a system's goal-directedness sometimes depends on extrinsic human activity, for example when a homing missile is launched. It is also not to deny that a goal-directed system is typically *causally* dependent on human aims, as when a programmer programmes a homing missile.

Core behaviourist concepts appealed to in this respect are the cybernetic notions of plasticity and persistence, which we discuss in Section 3.2. However, as we shall see, in order to overcome certain counterexamples, the cybernetic theory also requires some mechanistic constraints; and even this may not be enough. Later on, we shall argue that in order to deliver an adequate theory of goal-directed behaviour, we may need to posit irreducibly teleological properties after all. Hence, it is far from clear that Nagel's reductive ambitions can be achieved – at least where teleology is concerned.

3.2 Nagel's 'System-Property' Cybernetic Analysis

For the time being, let us see what prominent cybernetic theorists have had to say about our key questions: What are the defining characteristics of behaviour that is goal-directed? And, in truthmaking terms, what does the world have to be like in order for it to be true that certain behaviour is goal-directed? What is striking about cybernetic systems is that they can adapt their behaviour in response to various interferences, in such a way that they can still achieve their goal – or at least remain on track to achieve their goal. We find this same adaptability in other kinds of teleological system. For example, the human body reacts in various ways to both low and high external temperatures in order to maintain an internal bodily temperature of around 37 degrees Celsius. The similarities of this human subsystem with the thermostat example are striking, albeit the thermostat's mechanisms are much simpler. How, then, can this adaptability be spelt out in behavioural, empirical terms?

The cybernetic answer employs the concepts of plasticity and persistence, which were developed in the 1940s and 1950s by philosophers of biology such as E. S. Russell (1945), Sommerhoff (1950), Braithwaite (1953), and later Nagel (1961, 1979: Ch. 12). Roughly, a system is plastic if its goal(s) can be reached via different causal paths or from different initial positions (Nagel 1979: 286). As Garson notes (2016: 23), for theorists like Braithwaite, 'initial positions' does not merely mean different positions in a given environment but rather different environmental conditions altogether. As for persistence, this is more to do with the adaptability of the behaviour as it occurs, such that if one path to the goal is blocked, an alternative causal route to the goal is taken. As a system operates, any number of changes could occur – both in the external environment or within the system itself – which would prevent it from achieving its goal were its behaviour not modified. A system is persistent, then, if it is able to compensate for a range of such changes by modifying its behaviour in a way that still leads it towards the relevant goal (Nagel 1961: 411). One way of thinking about this idea is that the achievement of the goal by the system is to some extent independent of specific internal and external variations, at least within a certain range.

Notice that, on the surface, the concepts of plasticity and persistence are empirical and therefore testable. That is, we can test for plasticity and persistence by placing a system in different situations and seeing if its outward behaviour is modified accordingly in order to reach some hypothesized goal(s). If it is, then we have prima facie evidence regarding the goal of the system, which can then be confirmed through further testing (see e.g., Rosenblueth and Wiener 1950: 325; Beckner 1959: 143–144). This also means we will often be able to make judgements about plasticity and persistence while knowing little about the precise internal mechanisms of a system. Hence, we can, in principle, identify the goals of a system prior to knowing about the specific functions of its parts. This is an important feature to highlight because it avoids a potential worry about epistemic circularity. If the goal-contribution theory of functions is correct, then knowledge about the functions of things depends on knowledge about how those things contribute to the goals of a given system. However, we would be caught in an epistemic circle if it were also the case that we have to identify the functions of things before we can determine the goals of relevant systems.

How, then, can we construct a more rigorous metaphysical analysis of a goal-directed system? Nagel's own analysis is based on several formal elements (1961: 411–18): S is the system in question, E is the external environment, and importantly G is some goal state that S either possesses or is at least capable of possessing under certain conditions. Note that the goal state is not merely accidental but rather 'distinctive' of the integrated system in question (Nagel 1961: 422; see also McLaughlin 2001: 76). This implies that directively organized systems are individuated at least in part by their goal-types.

Next, we specify the components of the system which are causally relevant for the occurrence of the goal state, call them A, B, C .... There may be any number of such components. The states of such components are then represented by state variables 'A_x', 'B_y', 'C_z' and so on. These may or may not be numerical variables, depending on whether the states in question are quantitative or qualitative. The variables are such that they can change, and the overall state of S at any given time is expressed by the matrix $(A_x B_y C_z \ldots)$. Nagel also stipulates that the various state variables must be suitably independent of each other, or 'orthogonal', but we may set that idea aside for now.

Suppose that S is in its G state at t_0 but is such that a change in any one of the state variables A_x, B_y, $C_z \ldots$ would take S out of G (assuming that the values of the other variables remain the same). Following Nagel, call such a change a 'primary variation'. Then, we can say that S is directively organized or goal-directed if S's parts A, B, C ... are such that if primary variation occurs, the states of the relevant components change in such a way that at some later time t_1 the system is still in state G (or perhaps tending towards G). In short, the idea is that when primary variation occurs, a directively organized system's parts adapt and compensate in such a way that G tends to be maintained. This is just a more formal way of capturing the ideas of plasticity and persistence. One of the features of this formal definition is that it remains silent about the precise mechanisms which give rise to the compensatory behaviours (Nagel 1961: 418). With this basic definition of directive organization in play, we can then add (as appropriate) the stipulation that S's state variables also compensate for relevant *environmental changes* in such a way that G is maintained (or is on course to be maintained). To do so, we merely require a state variable for E ('F_w') as a further possible source of variation.

At this point, one might wonder just how many potential changes a system needs to be able to compensate for in order to be regarded as plastic and persistent. This is a delicate question because to put any specific number on it would seem somewhat arbitrary. The answer given by Rosenblueth and Wiener (1950), Braithwaite (1953), and Nagel (1961) is that plasticity and persistence come in *degrees*. This idea has been developed more recently by Babcock (2023), Lee and McShea (2020), and McShea (2012). In Nagel's analysis, a directively organized system compensates for so-called primary variations in order to maintain some goal state G. However, the range of primary variations that a system can compensate for – call it K_A' – may vary between cases. Hence, systems can be more or less directively organized depending on how inclusive the range K_A' is (Nagel 1961: 417). We shall return to this issue in Section 4.7.

With all these details in place, we can reconstruct the Nagelian theory roughly as follows, starting with the notion of directive organization ('DO'):

DO: A system S is directively organized towards goal G if and only if G is a global property of S by which S is (at least partly) individuated, and G is achieved though systemic behaviour that is plastic and persistent, i.e., adaptable in the face of a range of primary variations.

Recall that the goal-contribution analysis of function performance ('GCF'), which Nagel also appears to endorse, is as follows:

GCF: X performs function Y for system S at t if and only if the Y-ing of X at t makes a causal contribution to a goal G of S.

Importantly, with these definitions in play, it seems we can generate non-teleological equivalents of function statements and functional explanations. When we explain something's behaviour functionally, we are merely explaining

Teleology 31

what contributions the behaviour makes to the plastic and persistent behaviour of some complex system. And for the Nagelian reductionist, plasticity and persistence are defined in non-teleological behaviouristic terms. Hence, although Nagel is a goal-contribution theorist about functions, it turns out that, for him, talk of goal-contribution (and thus functions) is a convenient heuristic that can ultimately be explained away. This suggests that the cybernetic theory of goal-directedness is compatible with purely mechanistic worldviews, such as those inspired by Descartes and Hume, where the only kind of causation at work is so-called efficient causation. According to the Humean metaphysics, the world is just a mosaic of loose and separate qualities which happen to form certain patterns (see Williams 2019: Chs. 1 and 2). On this view, these patterns are brute facts which are not explained by any deeper metaphysical principles. This kind of picture stands in stark contrast to the anti-Humean theory of goal-directedness proposed in Section 4, which appeals to irreducible, end-directed powers.

Assuming for now that the Nagelian reductive analysis is feasible, why do the explanatory practices of biologists and other special scientists often favour teleological formulations? This is a compulsory question, because even Nagel accepts that teleological explanations are pervasive in some areas of science and are often taken to be indispensable. Nagel's answer is that whether one deploys a teleological or a purely mechanistic explanation is context dependent and largely to do with emphasis or selective attention. Explanations involving goals are forward-looking, placing an emphasis on the end point of a systemic process to which various parts contribute. In contrast, mechanistic explanations are more backward-looking, focusing on the conditions that give rise to the relevant process (Nagel 1961: 405, 421–422; see also Schaffner 1993: 391).

With the core elements of this reductive strategy in place, let us discuss some problems facing the theory.[22]

3.3 The Problem of Extensional Adequacy

As with most philosophical analyses purporting to provide necessary and sufficient conditions for some phenomenon, critics have promptly replied with alleged counterexamples. In response to the alleged counterexamples, cybernetic theorists have either bitten the bullet and accepted the (perhaps counterintuitive) consequences of the analysis, or modified the analysis accordingly. In this section we shall outline some of the moves that cybernetic theorists

---

[22] I shall not attempt to discuss all possible criticisms here. I recommend Cartwright 1986 for a sophisticated scientific critique of reductionism about teleology. For critiques in the phenomenological tradition, see for example, Coyne 2016, Gambarotto 2020, Jonas 2001, and Sachs 2023.

have made in response to various alleged counterexamples. However, since such counterexamples tend to rely on intuitions about what does and does not fall under a concept, it is difficult to say whether firm conclusions can be drawn from this area of the debate. For this reason, I shall not dwell on this issue for too long.

The counterexamples that cybernetic theorists have taken most seriously are those which threaten the *sufficiency*[23] of the plasticity and persistence conditions for goal-directed behaviour, suggesting that the analysis is guilty of overbreadth. Nagel (1961, 1979) discusses several examples of relatively simple processes which appear to display plastic and persistent behaviour.[24] One example concerns a ball at the bottom of a bowl that is disturbed in various ways but always behaves in such a way that it returns to its initial position (1979: 288). Another example involves a pendulum (Taylor 1950a: 316, Nagel 1961: 419–420) which is disturbed by a gust of wind but returns to its lowest point via gradually decreasing oscillations. What these examples have in common is that some equilibrium state is restored in the face of disruption. Such processes are familiar in physics and Bedau (1992b) discusses a range of similar examples, arguing that we would not regard them as cases displaying teleology.

Faced with such examples, cybernetic theorists have typically been reluctant to bite the bullet and accept that such systems are goal directed. Instead, they have added more necessary conditions for goal-directedness in order to preclude those cases from counting as goal directed. It is not obvious that this is the correct strategy, however. As noted earlier, Braithwaite and Nagel accept that directive organization comes in degrees: one system can be more or less plastic and persistent than another. With this in mind, one wonders why we shouldn't just accept that the ball in a bowl *is* goal-directed but to a lesser *degree* than, say, a homing missile (see e.g. McShea 2012; Babcock 2023: Sect. 3).[25] If one takes the ball out of the bowl, then of course it will not modify its behaviour in such a way that it travels back into the bottom of the bowl. Nonetheless, there is a relatively narrow range of circumstances, within the bowl, in which the ball persistently makes its way back to its starting position.

---

[23] The *necessity* of plasticity and persistence for goal-directed behaviour has also been questioned on the basis of examples suggested by Woodfield (1976: 98–99, 191) and Ehring (1984b: 218). See their cuttlefish and lottery examples, respectively.

[24] Relatedly, so-called 'extremal' principles in physics are often thought to give rise to activity that is quasi-teleological (Nagel 1961: 407). Such principles, like the principle of least action, say that certain systems develop in such a way as to minimize or maximize some magnitude. See also Hawthorne and Nolan 2006.

[25] I suspect that Rosenblueth and Wiener (1950: 319) would also agree with this strategy, since they seem happy to accept that relatively simple processes are goal-directed, citing as examples a weighted roulette wheel and a magnetic compass.

This strategy would still accommodate the intuition that homing missiles and renal systems are the more paradigmatic examples of goal-directedness, on the basis that they are goal directed to a much higher degree. I leave readers to judge the merits of that strategy. For completeness, I shall also briefly outline the sorts of modifications that Nagel and others have made in response to these perceived counterexamples.

As Garson puts it (2016: 23), in order to preclude pendulums and balls-in-bowls from counting as goal-directed systems, the cybernetic theories have taken an overtly 'mechanistic' turn. What Garson means is that cybernetic theorists have taken a closer look at the internal mechanisms of goal-directed systems, with a view to showing that pendulums and balls do not have the right sorts of inner organizational complexity to count as being goal-directed. A typical ball, for example, is 'internally homogenous' (2016: 23), whereas paradigmatic goal-directed systems have clearly delineated parts which make distinctive contributions in a coordinated effort towards a goal.

In the philosophical literature, there are at least two different ways of spelling out this further condition of organizational complexity, though the two are arguably compatible and complementary. The Nagelian condition, which has its roots in the work of Sommerhoff (1950), is a formal condition requiring that the states of the parts of a directively organized system are *independent*, in the sense that the values of the state variables they instantiate must not determine the values instantiated by other parts of the system at that same time. In Nagel's terminology, the different variables must be 'orthogonal' (1979: 288). Consider the following example from Nagel (1979: 289): the function of a Watt governor in a steam engine is to maintain speed (within a certain range) by controlling the amount of steam entering the engine cylinders. As with other negative feedback mechanisms, the system involving the engine-plus-governor is plastic and persistent. And, importantly for current purposes, if the 'Watts governor of a steam engine is not hitched up to the engine, any speed of the engine is compatible with any speed of the arms of the governor; for there are no known laws of nature according to which, in the assumed circumstances, the spread of the arms depends on the engine speed' (Nagel 1979: 289).

Since the Watt governor system is a goal-directed system *par excellence*, this raises the prospect that independence of the relevant variables is a necessary condition for goal-directedness. Importantly, this condition appears to preclude pendulums and balls-in-bowls from being goal directed. In the ball case, the relevant variables that determine the ball's motion, that is, those concerning restoring and displacement forces, are not independent of each other because they are both fixed by the laws of motion.

The other way of imposing an appropriate mechanistic constraint is to insist that goal-directed systems involve *negative feedback* mechanisms. The Watt governor is, like a thermostat, an example of a negative feedback mechanism – arguably one of the first. Rosenblueth et al. (1943) were arguably the first to offer a negative feedback analysis of goal-directed behaviour, which was taken up subsequently by Adams (1979) and Faber (1986). How, then, are negative feedback mechanisms best characterized? As one would expect from a plastic and persistent system, a feedback mechanism exhibits complex behaviour (the output) in response to an input. Hence, such systems have something resembling a detector or sensor, which allows them to detect the values of relevant variables so that the output can be adjusted accordingly. However, in a *feedback* system there is also a causal loop. As Rosenblueth et al. (1943: 19) put it, in feedback systems 'some of the output energy of an apparatus or machine is returned as input', resulting in a continuous cycle of output and input. In cases of *negative* feedback, the input can be used to restrict rather than amplify the output, ensuring that a certain goal-threshold is not exceeded. If such systems can react accordingly to a range of inputs, then such systems will naturally satisfy the behaviouristic criteria of plasticity and persistence discussed earlier. Homeostatic mechanisms such as thermostats provide good examples: as Taylor nicely puts it, 'a thermostat controls, and is in turn controlled by, temperature' (1950a: 315). In more complex examples such as that of a self-driving car, the system continually monitors the environmental conditions and adjusts its behaviour accordingly, typically through regression algorithm computation.

Clearly, pendulums and balls-in-bowls are not negative feedback mechanisms, so the negative feedback requirement also avoids the salient counterexamples. Moreover, I take this strategy and Nagel's independence condition to be complementary, for it is difficult to think of an example of a negative feedback system whose controlling variables are not nomically independent. So, one way of thinking about the negative feedback condition is that it specifies a mechanism which realizes Nagel's formal independence condition. But whether other mechanisms could also perform this role is an interesting question that we shall not address here.

I leave readers to judge the merits of these modifications.[26] But regardless of their merits, the reductive cybernetic analyses still face problems that are more fundamental, namely the problems related to goal failure and underdetermination. These objections stem from the behaviouristic roots of the cybernetic theory, and it is to those problems that we now turn.

---

[26] For further critical discussion of Nagel's independence condition, see Bedau 1992b and Garson 2016: 30.

3.4 The Big Bad Bug: The Problems of Goal Failure and Underdetermination

In its basic form, the cybernetic view is based upon patterns of plastic and persistent behaviour, and therefore assumes that the goal states of a system are typically achieved. For something to demonstrate goal-directed behaviour, it must show that it can follow more than one causal pathway *to the goal*, and also modify its behaviour so that it still reaches its goal. As Nissen puts it, the cybernetic theory of Braithwaite and others is based on trial and success (1997: 7). This leads to what Nissen calls the *problem of goal failure*.

The problem is that it is overwhelmingly plausible to think there can be goal-directed behaviour even if the relevant goal is never achieved. To use one of our stock examples, a homing missile might miss its target if the target emits sophisticated decoy countermeasures. But even in that case, it still seems correct to say that the missile's goal was to destroy the target. There can also be goal-directed behaviour in cases where the object of the goal *never* exists, as when scientists search to discover a certain element without hope of success. In such cases, it is difficult to see how the condition of plasticity can be satisfied, for if the goal never exists, there cannot be alternative causal routes to it. This problem looks especially serious for those theories that include the negative feedback condition. As Nissen (1997: 30) and Garson (2016: 27) remark, it is arguable that *by definition*, negative feedback systems are ones which modify their behaviour by receiving input from their target. In that case, it is surely impossible for a negative feedback system to exhibit plasticity and persistence if the target does not exist.

Cases of goal failure or missing goals were identified as a problem for behaviourist analyses of goal-directedness as long ago as 1950 (Taylor 1950b: 329), and have since been elaborated by, amongst others, Scheffler (1959, 1963), Beckner (1959), Hull (1973), Ehring (1984a), and Nissen (1997). To deal with such cases, goal theorists must surely avoid any implication that a goal-directed system has to achieve its goals. However, this is not a straightforward task for the cybernetic theorists. As soon as they accept goal-directedness in cases where the goal is never achieved, they are left with a problem of *underdetermination*. In its basic form, the cybernetic theory grounds goal-directedness in patterns of behaviour, but in cases of goal failure the behaviour that occurs does not lead to the achievement of a goal. But if a goal does not come about, then who is to say what it is that the behaviour is directed towards? For example, I imagine, sadly, that there have been times at night when our baby son has cried, but we haven't attended to him due to being in deep sleep. Who is to say what the goal of the crying was in those cases? Was the goal to acquire milk, to play, to be cuddled, or ...? Note that this underdetermination problem is not merely epistemic; for the cybernetic theorists it is also *metaphysical*. Intuitively, there should be a determinate fact of the matter as to what the goal is in such cases, but within a behaviouristic metaphysics it is not easy to see how there can be such a fact.

One is tempted, of course, to say that the goal in my example is determined by my son's *intention* to, say, be fed. But this solution is not an option for cybernetic theorists because they typically reject intentional theories of goals and functions. For one thing, it is far from clear that the concept of intentionality is non-teleological and hence admissible in an analysis of teleology that purports to be reductive. And even setting that issue aside, remember that the theory is meant to be applicable to inorganic cybernetic systems, which do not seem to be the sorts of things that can exhibit intentionality, except in a mere metaphorical sense.

Another option would be to ground the goals of cybernetic systems extrinsically in the purposes for which they are designed or used by human agents. But, again, this solution does not sit well with the cybernetic approach. As Taylor acknowledges (1950a: 312), Rosenblueth et al. are clear that cybernetic systems are *intrinsically* purposeful (1943: 19). Moreover, if the goals of cybernetic systems were determined extrinsically by the intentions of their human designers, this would push the cybernetic theory towards the kind of selectionist analysis of teleology discussed in Section 2. Rosenblueth and Wiener (1950) are sceptical of that approach in the case of goals, partly because they think that the goal-directedness of a machine could clearly differ from what its designer intended. For example, a complex weapon that we design could end up being disposed to kill unintended targets in a plastic and persistent way (1950: 318). Bigelow and Pargetter raise the same sort of problem for design-based theories of function, for it seems perfectly plausible that some things can regularly perform functions for which they were not originally designed and created (1987: 185).

In order to overcome the problem of underdetermination, two main options have been offered by the cybernetic theorists. The first option is a counterfactual strategy and the second involves positing properties of inner representation in the relevant systems, as a way to determine the relevant goals. We shall begin in the next subsection with the counterfactual strategy, before briefly discussing the representation strategy in Section 3.6. Although these solutions are consistent with the intrinsicality of goal-directedness, we shall see that both theories face serious problems.

3.5 The Counterfactual Formulation

The earlier definition of directive organization (Section 3.2) captures what (allegedly) happens in actual cases of goal-directed behaviour. One of the lessons of Section 3.4 is that we should not insist that a system is goal-directed only if it *actually* demonstrates plastic and persistent behaviour. However, even if a goal-directed system does not actually demonstrate such behaviour, one might expect that certain hypothetical truths or 'counterfactuals' will hold, namely, that *if* a range of internal or external primary variations were to occur, then the system *would* adapt its behaviour in a way that still leads towards the goal. That is to say, in other words, that the *modal profiles* of directively organized systems have a plastic and persistent structure. Nagel himself occasionally acknowledges this point.[27] When discussing the example of the human subsystem that moderates water content in the blood, Nagel expresses the goal-directedness in counterfactual terms: '*were* the blood inundated with water to a greater or lesser extent than was actually the case, the activity of the kidneys or of the muscles and skin *would* have been appropriately modified' (1979: 287). Perhaps, then, such counterfactuals could be put to work to determine the goals of a system in cases where the goal is not achieved. Even if something prevented a particular human body from moderating its water content, the goal of water moderation would nonetheless be grounded by relevant counterfactuals concerning the kidney, muscles, and skin.

Consider Scheffler's favoured example of a dog that is trapped in a cave and is pawing at the door (1963: 119). This looks like a case of goal-directed behaviour *par excellence*. It is easy enough to imagine that by pawing at the door, the dog is striving towards the goal of leaving the cave. But if the door is never opened and the goal is never achieved, in what sense can we say that the dog's behaviour is goal-directed? If we want to avoid appealing to the dog's intentions, a counterfactual solution looks like the obvious way to go (Woodfield 1976: 49). Even if the dog's goal is not in fact achieved, we might insist that the following counterfactual is still true of the dog: *if* the door had opened in various ways, then the dog would have exited. The idea, then, is that the consequents of such counterfactuals determine the goals towards which the systems are directed.

On the face of it, this looks like a neat and straightforward solution. However, this strategy faces at least two significant challenges. The first problem is raised by Nissen (1997: 8). As I interpret Nissen, the worry is that the counterfactuals might not always line up perfectly with what the goal really is. Even if the goal of the dog's pawing really is to be let out of the cave, there is no guarantee that, counterfactually, the dog would leave if the door were opened. For one thing, goals can change in unexpected ways due to internal or environmental factors. For example, suppose that outside the cave there is a deep freeze. After the door were opened, the dog might feel the bitter cold on its nose and decide that, after

---

[27] See also Sommerhoff (1950: Ch. 2), who defines 'directive correlation' and 'adaptation' in counterfactual terms. For a recent formal presentation of the subjunctive stability of persistence and plasticity in biology, see Stovall (2024).

all, it is better to stay in the cave. So, the counterfactual would deliver the wrong verdict, suggesting that the goal was to stay in the cave rather than exit.

There are technical modifications that the cybernetic theorists could attempt so that the counterfactuals deliver the correct verdicts. But it is not going to be easy. The wrinkle in the counterexample was that the dog's goal was changed at the last minute. So, the counterfactual theorist might insert the following sort of proviso: if the door were opened in some way or other, *and the dog's goal were not altered*, then the dog would exit the cave. However, clearly this will not do. Again, the aim of Nagel and other reductive theorists is to analyse talk of goals and goal-directedness in non-teleological terms, so that the analysans does not itself involve terms like 'goal' or related ones such as 'purpose' or 'intention'. The problem with the proviso is that it deploys the very concept of goals that we are trying to understand.

Another option is to include a proviso in the counterfactuals' antecedents that appeals to 'ideal' or 'ordinary' conditions. In the example, perhaps the very cold weather outside the cave is just an unusual situation that the dog finds non-ideal. However, it might be urged that this is no counterexample because being goal-directed towards exiting the cave just means that, *if the circumstances were ideal*, the dog would leave the cave when the door were opened in one way or another. This solution is still not straightforward, however. In the debate about reductive counterfactual analyses of dispositions, the difficulty of spelling out 'ideal' provisos in an informative way is often highlighted (e.g., Martin 1994).[28] It would be unrealistic to offer a disjunctive list of all the possible ideal circumstances, for such a list would be incredibly long and impossible to formulate. And more worryingly, it seems we would already have to have some grip on what the dog's goal is in order to decide what does and does not count as an ideal circumstance. This is because the whole point of the 'ideal conditions' clause is that it captures just those circumstances in which the dog's goal is not altered or otherwise thwarted. So, one cannot help thinking that there is still circularity, albeit more covertly.

Even if a technical fix can be found which avoids circularity and is not too ad hoc, there is still a more fundamental and difficult metaphysical question to face. What exactly is it that determines the truth of the relevant counterfactuals? One should not let the truth of counterfactuals 'hang on air', as Armstrong would say (2004: 3), for that would render them mysterious. In the dog example, what is it about the dog or the world at large which determines how the dog would behave in the relevant counterfactual scenarios? The danger for

---

[28] For a more detailed discussion of the parallels between the debates about dispositions and goal-directedness, see Tugby's 2024 unpublished manuscript entitled 'The Property of Goal-Directedness: Lessons from the Dispositions Debate'.

reductionists like Nagel is that the answer will inevitably invoke teleological properties: it is precisely the goals or intentions of the dog that make it true that it would go through the door if it were opened in various ways. To put the worry in a slightly different way, the cybernetic theorists face a Euthyphro dilemma: do goals determine the relevant counterfactuals, or do the relevant counterfactuals determine the goals? For the cybernetic theory to succeed in its reductive aims, it needs to uphold the second disjunct in a way that does not presuppose teleological facts. I cannot rule out that there is some way to achieve this, but how to do so is, at the very least, far from obvious.

3.6 Inner Representational Properties

Other cybernetic theorists have employed a rather different strategy in order to address the sorts of problems raised in Section 3.4. This involves adding a new requirement, which is that a goal-directed system must contain some internal representation of the goal state (e.g., Adams 1979; Faber 1986; Manier 1971). In Adams's view, a negative feedback system operates not by receiving feedback signals from the target itself, but rather from a *representation* of the target. The idea, then, is that whenever the overt behaviour of a system underdetermines the goal, the goal is nonetheless fixed by an inner representation of the goal. For example, in the case of a thermostat, the target might be represented by the position of the temperature dial (Adams 1979: 507). Even if the target is not achieved, for whatever reason, the representational state of the feedback system determines the goal towards which it is directed.

To be clear, the claim that feedback systems like thermostats have inner representations is not to be interpreted merely as a metaphorical claim. Nor is it being claimed that goal-directed systems represent merely in the sense that minded creatures like us can interpret them in a certain way – as when we read a thermostat dial to learn which temperature it is helping to maintain. Again, the cybernetic theory typically regards the goal-directedness of, say, a homing missile as an intrinsic property of the missile. So, it had better be the case that the required inner representations are intrinsic to a goal-directed system.

With these clarifications in place, one might protest that such a view is clearly absurd, and insist that only organisms – and perhaps computers of a certain complexity – can literally have inner representations. Surely a thermostat is not complex enough to represent anything to itself, given that it doesn't have anything resembling a mind. Perhaps we could accept that talk of representation is appropriate in such cases in some sense, but only in the sense that we are employing a mentalistic metaphor or analogy (e.g., Woodfield 1976: 183, 194).

One way of developing this worry in a more precise way is to argue that representation requires *intentionality* and that many goal-directed machines clearly do not exhibit intentional states.[29] One reason for thinking that intentionality is required for representation is precisely that representational states can often represent things that do not exist. As we have seen, the cybernetic theory must be able to accommodate cases of goal-directedness towards missing goals, and in those cases it looks as though the object of the relevant representation will have to be an intentional object. However, this point again threatens the reductive ambitions of the cybernetic theory, because it is far from clear that the concept of intentionality is non-teleological. Moreover, intentionality is often regarded as the mark of the *mental* (Brentano 2015/1874). Again, surely it cannot be claimed with much plausibility that simple feedback mechanisms like thermostats literally have something resembling a mind.

One way of responding to this worry is to accept that goal-directed systems exhibit intentionality but deny Brentano's thesis that intentionality is the mark of the mental. That is, one could try to water down the notion of intentionality in some way and accept that cybernetic machines such as thermostats are in *physical* (i.e., non-mental) states of intentionality. How could one motivate such a claim? Here we might turn to a theory developed by Molnar (2003) that says that many physical properties are causal *powers* and that power instantiations have the marks of intentionality. This is an interesting and underexplored proposal, but unfortunately, as we shall see in Section 4.2, Molnar's intentionality account of physical powers faces serious problems. Again, there remains a feeling that insofar as we can ascribe states of intentionality to physical, non-minded things, we are doing little more than employing a mentalistic analogy or metaphor. But if that is all we are doing, then cybernetic machines cannot *really* be said to have inner properties of intentionality. And hence, as far as our theory of teleology is concerned, it remains far from clear which features of a cybernetic system ground its goal-directedness, especially in cases of goal failure.

3.7 Taking Stock

In this section, we have examined the reductive cybernetic theory of goals, which promises to explain goal-directedness and function attributions in entirely non-teleological, mechanistic terms. Because of its behaviourist roots, the theory faces the serious difficulty of accommodating and explaining goal-directedness in cases where the relevant goals are not (or even cannot) be

---

[29] For further worries about the inner representation strategy, see Ehring (1984b: 218–220), who argues that it generates circularity problems.

achieved. We have explored two possible strategies for overcoming this problem, a counterfactual approach and one based on properties of inner representation. Both of these strategies face serious difficulties. As things stand, therefore, we are still in need of a theory of goal-directedness that is able to provide an adequate metaphysical foundation for the goal-contribution account of functions. In the fourth and final section, we explore a novel non-reductive account according to which many cases of goal-directedness are grounded in certain end-directed *powers* of a system. According to this alternative view, while certain behavioural patterns and counterfactual truths might well be *evidence* of goal-directedness, they are not *constitutive* of it.

4 Powers
4.1 The Metaphysics of Properties: Powerful or Categorical?

So far, we have examined two related teleological phenomena that are familiar in many branches of science, namely, *functions* and *goal-directedness*. In this section, we shall investigate another related concept which is arguably teleological: the concept of *power*. The metaphysical concept of power has Aristotelian roots and has attracted much interest in contemporary philosophy. However, there is still much work to be done when it comes to understanding the connections between powers and other teleological phenomena such as that of goal-directedness. Later on, we shall explore the possibility that, in at least some cases, goal-directedness is grounded in certain *powers* that are exhibited by directively organized systems. I shall not attempt to argue conclusively in favour of the powers metaphysics, but what I do hope to show is that the powers theory provides resources for overcoming problems discussed in the previous section regarding goal-directedness. Let us begin by introducing realism about powers, exploring its teleological implications, and explaining why powers are important in contemporary metaphysics of science.

The best way to introduce realism about powers is by contrast with its main rival: categoricalism. Categoricalism and the powers theory are opposing metaphysical theories about the *properties* of individuals, such as the property of mass that is exemplified by all physical bodies. Other examples of properties include electric charge, molecular structure, biological traits, and the colours of a flag. All individuals have properties and typically many of them. According to categoricalism, properties have a primitive *non-dispositional* essence. Although categorical properties determine how things are, they do not in themselves dictate *what things can do*. For the categoricalists, worldly individuals are thus inherently inert and require the laws of nature to animate them. And such laws are typically regarded by categoricalists as being wholly contingent and external to individuals.[30]

In contrast, powers theorists claim that the relationship between an individual's properties and its behavioural dispositions is much more intimate. In their view, the connection is one of *metaphysical necessity*: if an individual has certain properties, then it cannot fail to have certain behavioural dispositions. It is thus properties, and properties alone, that explain the dispositions of things and the counterfactual truths associated with those dispositions. For powers theorists, individuals are inherently powerful by virtue of the properties that they instantiate;[31] we do not need external laws of nature to animate a world that is otherwise static and inert.[32] This is taken to be a significant benefit of the powers theory, because categoricalist theories of laws arguably face serious problems.[33] Moreover, the very notion of a categorical property strikes many metaphysicians of science as being obscure. We gain knowledge about the natural world through causal engagement with it, but if the categorical properties of individuals are wholly distinct from their causal dispositions, it is far from clear how we could ever gain knowledge about categorical properties.

What, though, does it mean to say that a property is inherently powerful rather than categorical? And why do many powers theorists regard powers as teleological entities? These are questions that we shall probe in the following two subsections, before exploring possible connections between powers and goal-directedness. An important point to acknowledge for now is that power properties are characterized, at least in part, by the end-states or 'manifestations' that they are powers *for*. Thus, when explaining powers, considerable emphasis is placed on the *directedness* that a power has towards its manifestation(s). Indeed, it is this directedness which distinguishes a power-type from others. For example, if someone wished to know about the powers associated with a certain mass property, we would naturally highlight the gravitational and inertial behaviours towards which the relevant masses are disposed. So, it seems that the identity of a power is determined, at least in part, by a certain end (or ends) towards which the power is directed. Such

---

[30] For a good overview of categoricalist conceptions of laws (amongst others), see Hildebrand (2023).

[31] There are different fine-grained views about the metaphysical nature of dispositional properties. Some think that all properties are simultaneously dispositional *and* qualitative (e.g., Giannotti 2021; Martin 1993) while others think that dispositions are necessarily grounded in qualities (e.g., Kimpton-Nye 2021; Tugby 2021). Perhaps the most common powers theory is that the nature of a property is exhausted by the dispositions it confers: properties are 'pure powers' (e.g., Bird 2007; Ellis 2001; Mumford 2004).

[32] Mumford (2004) goes as far as to say that powers theorists can eliminate laws of nature altogether. But see Bird (2007: Ch. 9) for a more moderate approach.

[33] For accessible discussions of the problems, see Mumford (2004) and Hildebrand (2023).

characterizations are *forward-looking* and therefore appear to be teleological in some sense.

George Molnar (2003) was one of the first powers theorists to put the directedness of all powers at centre stage, but as we shall see in Section 4.2, Molnar arguably takes the idea too far when he characterizes power directedness as a species of *intentionality*. In Section 4.3, we shall explore a more promising proposal, which is that powers are primitive telic states. If the powers theory is correct, the world is arguably teleological with a big 'T'.

4.2 The Intentionality Theory of Powers

Dispositions and their directedness were traditionally analysed in terms of counterfactual conditionals: for example, some would say that fragility is directed towards breakage in virtue of the fact that *if* a fragile object were struck with suitable force, then it would break. However, Molnar (2003) and others raised many serious problems and counterexamples for counterfactual analyses of dispositions (see Friend and Kimpton-Nye 2023 and Schrenk 2016: Ch. 2 for historical summaries). Although disposition ascriptions might well *entail* certain counterfactuals, the prospects for *analysing* dispositions in wholly counterfactual terms appear dim. This has led to a resurgence of the idea that dispositions and their directedness cannot be analysed away in the way that neo-Humean reductionists hope. Dispositions and their directedness might well be irreducible features of the world.

This philosophical background provides the context for Molnar's work on powers, which sees the need for a new approach to dispositions. It is one that aligns dispositions with intentionality rather than counterfactuals. Notice that when creatures like us have thoughts, we invariably focus in on objects. That is to say, our thoughts are *about* and *directed at* objects. The essential *aboutness* of thought is what philosophers call 'intentionality'. Given that intentional states are directed at certain objects, Molnar was led to consider whether powers just are states of so-called *physical intentionality*.

The parallels between dispositions and mental intentionality had been highlighted earlier by Martin and Pfeifer (1986) and Place (1996). The intentional theory of powers is prima facie appealing because, like intentional states, dispositions are directed entities and can (it seems) be directed towards non-existent outcomes. For example, solubility is directed towards dissolving, even if soluble substances are never placed in circumstances where dissolving occurs. Similarly, in the case of mental intentionality, one can have a belief that is directed at unicorns even if there are no unicorns. Nonetheless, despite these surface similarities, it seems fair to say that Molnar's intentionality proposal is radical. Since fundamental physical entities such as electrons have powers like charge and spin, Molnar has to say that even electrons have intentional states. On the assumption that electrons do not exhibit mental properties, Molnar's theory implies that Brentano (2015/1874) was mistaken when he declared that intentionality is the mark of the mental: rather, for Molnar, intentionality is the mark of *power*.[34]

For all its attractions, it seems fair to say that the intentionality theory of powers has gained few adherents.[35] Most powers theorists would agree that there are some loose similarities between powers and intentionality, and in what follows we shall acknowledge some of those similarities. However, there remains a feeling that Place and Molnar take a step too far in claiming that powers literally are intentional states. Mumford (1999), Bird (2007), Oderberg (2017), and Marmodoro (2022), among others, have put forward several objections, many of which look strong.

A preliminary worry is that the explanations commonly given for the existence of intentionality are inappropriate in the case of physical powers. For example, many think that intentionality and representation are intimately connected: a thought can be about a non-existent object precisely because of the representational capacities of the thinker (e.g. Kroll 2017: 32; Marmodoro 2022: 7).[36] However, we saw in Section 3.6 that it is not easy to make sense of the idea that inanimate systems like thermostats have inner representations, not to mention simple entities like electrons.

There are, however, other general features that intentionality is commonly taken to have. Bird (2007: 119–120) lists a number of such features, which include the following:

i) Directedness towards an object: for example, a thought about George Molnar.
ii) The intentional object may not exist: for example, a thought about a unicorn.
iii) Intensionality: for example, it might be true that Sapna believes George Orwell wrote *1984*, but false that Sapna believes Eric Blair wrote *1984*, even though 'George Orwell' and 'Eric Blair' are co-referential.
iv) Indeterminacy of intentional objects: for example, a thought about the pint pot being somewhere on the bar but not anywhere in particular.

---

[34] Alternatively, someone like Molnar could agree with Brentano's account of the mental and accept the panpsychist thesis that all properties are mental. I take it that this is not Molnar's preferred option, but for further discussion of panpsychism, see Goff 2017: Ch. 7.

[35] Though see Bauer (2022) for a recent development of the intentionality approach to powers. Unfortunately, this Element was near completion when I discovered the Bauer source, and I am unable to give it the attention it deserves here.

[36] For further discussion about the role of representation in intentionality, see Crane (2008), Nes (2008), and Raimondi (2021).

Now, Bird and others are happy to concede that powers have the characteristics i) and ii), suitably understood. However, it is far from clear that any firm conclusions can be drawn from this regarding the connection between powers and intentionality. In many cases the directedness involved in powers looks rather different to that which is typically involved in mental intentionality. In Brentano's work, the directedness of intentionality is first and foremost an *aboutness*. But as Marmodoro rightly notes (2022: 7–8), the directedness of powers is more *dynamic*: powers strive causally to bring about their manifestations when activated. In contrast, there is nothing inherently dynamic about intentional aboutness, for we can have a belief about an object without striving for that object. So it is far from clear that the directedness of intentionality is relevant for understanding the nature of powers. These observations feed into a more general point, which is that directedness can occur in all manner of ways. For example, an arrow can be directed towards a target but not in the sense of intentional aboutness (Mumford 1999: 221).

What about the other marks of intentionality? Bird argues that they fail to hold in the case of powers. I shall not go through all of Bird's arguments here, but the indeterminacy condition (iv) is worth a closer look because here one might think that Molnar is on stronger ground.

Thoughts can be indeterminate in the sense that we can abstract away from certain details concerning intentional objects: we can think about a pint pot being somewhere on the bar without thinking of it being anywhere *in particular* on the bar (see e.g., Anscombe 1968: 161). Is there a similar phenomenon in the case of powers? Here, Molnar appeals to powers that are probabilistic, namely, propensities. Science suggests that the world may well be indeterministic, at least at certain scales. When an electron is released in the double slit experiment, there is no way of predicting exactly where it will end up. This suggests that the electron's power for motion is a propensity rather than a sure-fire disposition. In such cases, one might think there is vagueness regarding what the eventual manifestation will be, in the same way that there can be vagueness about where on the bar the pint pot is located in one's thought about it. So there is arguably some kind of parallel here. However, there are reasons to think that the analogy is not close enough to show that powers and intentionality are the same kind of phenomenon – especially given all the other worries that have been raised about Molnar's proposal.

The reason why the analogy is only rough is that the main source of indeterminacy in the case of thought is that of *abstraction* (see e.g., Oderberg 2017: Sect. 6). Details of where exactly the pint pot is located on the bar could in principle be filled in, but when thinking of the pint pot we choose to ignore unnecessary details and abstract away from them. However, in the case of propensities, surely nothing like abstraction is taking place. Rather, the indeterminacy of propensities consists in the fact that any one of a number of determinate states might come about: propensities have a less-than-one chance of doing this or that, even in ideal circumstances. This means that the future behaviour of the propensity's possessor is not fully determined. Hence, the kind of indeterminacy here looks rather different to that in our pint pot example. In the latter example, it is not as though we are unsure as to which determinate location we are thinking about; rather, all such determinacy has been completely abstracted away. So, although there is perhaps a loose analogy here, inasmuch as both cases involve vagueness of some sort, the vagueness involved in each case is rather different and has a different source. As Bird puts it, one can't help thinking that Molnar is conflating in*determinacy* and in*determinism* (Bird 2007: 125).

I think we have already said enough to cast doubt on Molnar's intentionality account of dispositional directedness. Even though the account does justice to the directedness of powers, as well as the idea that powers can be directed towards manifestations that do not occur, on close inspection the directedness of powers appears to differ from intentionality in important respects. We shall now, in Section 4.3, examine Kroll's more recent alternative (2017), which is overtly teleological. In the subsections to follow, we shall then return to the topic of goal-directedness and see how the powers theory can help us to tackle the problem of goal failure.

4.3 Powers as Telic States

In the previous section, we have considered one attempt to explain the directedness of powers in terms of some other concept: that of intentionality. In the absence of such an explanation, some recent powers theorists have proposed that it is simply a primitive fact about powers that they are end-directed, teleological states (e.g., Koons and Pruss 2017; Kroll 2017; Oderberg 2017; Tugby 2020; Witt 2008). This idea is sometimes thought to rejuvenate the Aristotelian concept of final causation: powers act in the way that they do precisely because they are directed at certain end states rather than others (e.g., Giannini and Mumford 2021: 88; Oderberg 2017). The view we shall consider does not try to *explain* the directedness of powers in more basic terms; it thus regards the world as being irreducibly teleological with a big 'T'. Nonetheless, as we shall now see, we can gain understanding of this directedness through the theoretical roles that powers play.

In what follows, I focus on Kroll's recent teleological analysis, partly because it is one of the most detailed of its kind, and also because the analysis emphasizes the end-directedness of all dispositional states. To begin, a quick terminological point is in order: Kroll frames his discussion using the terminology of dispositions rather than powers; however, as we shall see, it seems clear that Kroll holds a powers-based view about dispositions. But unlike Molnar, Kroll spells out the nature of powers in terms of a primitive notion of end-directedness rather than intentional directedness, leading to an overtly teleological account of dispositions ('TAD'). Kroll helpfully summarises TAD with three theses labelled (T2), (T3), and (T4):

'**(T2)** Necessarily: a property *P* is a disposition iff there is a condition *C* and event-type *M* such that necessarily, *P* is the property of being in a state directed at the end that one *M*s when *C*.'

'**(T3)** Necessarily, the property of being disposed to *M* when *C just is* the property being in a state directed at the end that one *M*s when *C*.'

'**(T4)** If *x*'s disposition to *M* when *C* is activated, then *in virtue of x being in a state directed at the end that x M's when C*, either *x* immediately *M*s or there is a process directed at the end that *x M*s.' (2017: 20–21)

(T2) tells us what it takes for some property to count as dispositional, while (T3) is a metaphysical thesis telling us that each dispositional property is *identical* with a certain state of directedness. (T3) supports my judgement above that, on this theory, dispositions are *powers*, because the right-hand side of (T3) describes what looks very much like a power. As we have already seen, a power is commonly defined precisely as a state of directedness towards its manifestation, as in Molnar (2003, Ch. 3; see also Bird 2007; Lowe 2010; Tugby 2013). Indeed, Kroll is explicit that 'Following Molnar (2003), I claim that directedness is what sets dispositions apart from non-disposition properties' (2017: 20). In what follows, then, I shall continue to use the terms 'disposition' and 'power' interchangeably.[37]

The teleological part of the analysis is, of course, 'the state directed at the end that one *M*s when *C*'. We can understand this primitive teleological notion of end-directedness through some of its theoretical roles and implications. (T4) is one such implication: a state of directedness is something which, when activated, either ensures that the target manifestation occurs immediately, or else ensures a process occurs which is directed at the end that *x M*s.

---

[37] Readers should, however, be aware that these terms are not always used interchangeably. Some authors maintain that dispositions are context-sensitive whereas powers are not (e.g., Vetter 2015: 81 and Williams 2019: 55). Relatedly, some regard dispositions as mere 'predicatory' properties, as opposed to powers which are 'ontic' properties (e.g., Friend and Kimpton-Nye 2023). To be clear, we are here employing the term 'disposition' in a way that is consistent with an ontic, realist conception of dispositions.

This second disjunct of (T4) introduces a second primitive notion of teleological directedness that applies to processes rather than properties. Such processes are initiated when a power is activated, as when a soluble substance begins the process of dissolution.[38] Again, although the directedness of a process like dissolution is a teleological primitive, we can gain understanding of it through its theoretical roles. For example, Kroll proposes the following principle regarding teleological processes (2017: 19):

'**(TP)**: If a process p is directed at end E, then, *in virtue of p being directed at end E, if p were to continue without interruption, E would be the case*'.

Here, Kroll is drawing upon Makin's Aristotelian insight that 'a teleological process has a privileged stage to which it runs in normal conditions, unless interfered with or hindered: the [end] to which it is directed' (Makin 2006: 194, quoted in Kroll 2017: 19). Putting these ideas together, Kroll arrives at the following conditional truth about dispositions:

'**(CDC)**: Necessarily: if x is disposed to M when C, x's disposition to M when C is activated, and x doesn't immediately M, then there is some process such that:

(*) *if* the process were to continue without interruption, x would M' (Kroll 2017: 19).

Now, although this theory of dispositions is clearly teleological, Kroll does not himself employ talk of goals or goal-directedness in his paper. In line with standard dispositional terminology, Kroll describes the relevant ends as manifestations and describes teleological behaviour in terms of the *end-directedness* of powers or processes.[39] Nonetheless, since powers come in different shapes and sizes, it would not be surprising if this powers framework could also yield a concept of goal-directedness in certain cases. Our discussion of goal-directedness in Section 3 already provides clues about what a powers-based notion of goal-directedness might look like. We could think of goal-directedness as a special case of end-directedness that we find only at certain levels of complexity, where the relevant powers have a *plastic* and *persistent* modal profile. In such cases, we might speak of a directively organized system's ends as its *goals* and describe its directed behaviour as *goal-directed* behaviour.

---

[38] Examples of such processes can be drawn from all scientific domains, including processes that have been naturally selected. For an interesting example, see Bellazzi's discussion of biochemical functions (forthcoming), which are associated with sets of dispositions to contribute to biological processes that have undergone natural selection.

[39] In contrast, Cartwright (2019: Ch. 2) is happy to describe all powers as being goal-directed, in the sense that there are essential constraints on what kinds of outcome a power can and cannot influence.

Teleology

To return to a point made in Section 3, in mechanistic terms, goal-directed powers will be the kinds of powers had by negative feedback systems, where the output of the behaviour is returned as input as part of a regulatory, reciprocal process.[40] Notice also that by restricting talk of goal-directedness to the powers of directively organized systems, we would avoid the problem facing Cummins's causal role theory of over-generating functions.[41] A powers theorist could insist that it is only *goal-directed systems* that involve functions, even though all powers (systemic and non-systemic) are teleological. In other words, the powers theorist could draw a distinction between powers that are *merely* end-directed, such as the powers associated with the charge of an electron, and those plastic and persistent powers of complex systems which are *goal-directed*.

In the remaining subsections, we shall explore this powers-based account of goal-directedness in more detail and consider whether *all* cases of goal-directedness can be grounded in powers. We shall start, in Section 4.4, by saying more about the powers of directively organized systems and how such powers might provide a solution to the problem of goal failure.

4.4 Another Look at Goal-Directedness: A Powerful Approach

In Section 3, we introduced the notion of a directively organized, plastic and persistent system. Such systems have one or more 'goal' states, which are a proper subset of the possible overall states of the system. It is plausible that various types of directively organized system are individuated, at least in part, by the kinds of goal-directed behaviours that are distinctive of them (Nagel 1961: 422). This is especially evident in the case of feedback systems because such systems are defined, at least in part, by their inputs and outputs.[42] However, this does not mean that a system must always be *actively* engaged in its goal-directed behaviour in order to be regarded as a goal-directed system. Systems can be in a latent state, as when a thermostat is switched off. Indeed, it seems perfectly possible for a thermostat to forever remain switched off and yet still be regarded as a goal-directed system, providing it has the *potential* for goal-directed behaviour. Nagel's counterfactual formulation of the cybernetic theory, discussed in Section 3.5, accommodates this point, but we saw that it remained unclear what grounds the truth of such counterfactuals.

It is here that the metaphysics of powers can be put to work because the powers theory insists that counterfactuals are grounded in dispositional properties, rather than vice versa. This theory maintains an intimate connection

---

[40] There are interesting connections here with Martin's concept of 'use' (2008: Ch. 9).
[41] I am grateful to Giacomo Giannini for emphasizing this point.
[42] This is arguably a modern application of the Aristotelian concept of *ergon*, i.e., a 'characteristic activity' (McLaughlin 2001: 76).

between dispositions and counterfactuals while resisting the tradition of reducing the former to the latter (as in Lewis 1997). Within this dispositional realist framework, we may regard goal-directed systems as having *powers* to bring about the relevant goals. We might then think of artefacts like thermostats as having an irreducible *dispositional* character or essence. This idea reflects Mumford's point that '[t]hermometer, light switch, computer, bookcase, towel rail, engine, and door handle can all reasonably be called dispositional terms' (1998: 197).

If we maintain that the characteristic powers of a directively organized system determine its goals and guide its goal-directed behaviour, it seems we have a firm theoretical foundation for talk of goal-directedness in at least some cases. Here we can take broad inspiration from recent authors such as Anjum and Mumford (2018: Chs. 2 and 9), Austin and Marmodoro (2017), Cartwright (2019, Ch. 2), Feser (2009, 2014), Koons and Pruss (2017), Kroll (2017), Oderberg (2017, 2020), Page (2015, 2021), Paolini Paoletti (2021b), Tugby (2020), and Witt (2003, 2008), who all see that realist dispositions or powers are intertwined with various teleological notions. However, as far as I know, few authors have put the powers metaphysics to work in the specific case of goal-directed systems. Importantly, if it can be shown that the metaphysics of powers can deliver an account of goal-directedness, a new and neglected advantage of the powers theory will be revealed.

Some clarifications might be helpful before proceeding. First, the proposal is not that goal-directed systems have *only* a dispositional character. Some systems might have a morphological nature too. The human blood circulation system is often called the 'cardiovascular system', conveying the idea that such a system must at least have a heart and blood vessels as anatomical parts. Nonetheless, it is clear that if some type of biological system did not generally have the disposition to pump blood, then it would not be regarded as a circulatory system. On the other hand, in some cases physical structure seems to play little, if any, role when it comes to individuating a system. This partly reflects the fact that many systems are multiply realizable: the same type of system, such as a thermostat, can be realized by different mechanisms, and we can often categorize systems while knowing little about their inner workings.[43] If a putative thermostat has the disposition to help regulate temperature in a certain way, then we can be pretty sure it really is a thermostat, regardless of what its inner structure turns out to be.

---

[43] Interestingly, this multiple realizability might suggest that many kinds are irreducible in some sense. Unfortunately, I do not have the space to discuss the ontological status of kinds here. But for related discussion see for example, Franklin and Knox 2018 and Bellazzi 2022 (I thank a referee for these suggestions).

Note also that although we are employing the notion of a disposition, our theory diverges in important ways from Cummins's theory of functions that we discussed in Section 2. Cummins's causal role theory is often regarded as providing a dispositional analysis of function. Nagel's reductive version of the goal-contribution theory of functions is also sometimes characterized in dispositional terms (as in Faber 1986: 61). However, it would be a mistake to think that the theory of goal-directedness explored in this section is aligned with the theories of Cummins or Nagel. Although, like Nagel, we favour a goal-contribution analysis of function (Section 2), the difference is that we are proposing an underlying metaphysical framework that involves a heavyweight notion of dispositionality, whereby dispositions are real, irreducible powers out there in the world. If we adopt a powers-based version of the goal-contribution account of functions, functions will be ascribable to a system's parts when they are able to contribute to the system's powers to produce the relevant goal state(s). Cummins's causal role theory and Nagel's version of the goal-contribution theory of function make no such metaphysical commitments to powers, and their mechanistic theories are compatible with reductionist or deflationist conceptions of dispositions, for example those on which disposition ascriptions are reducible to counterfactuals. Another way of putting this point is to highlight that such mechanistic theories are perfectly compatible with a neo-Humean metaphysics, on which causation supervenes entirely on the spatiotemporal arrangement of instantiated categorical properties. In contrast, the theory we are exploring is very much anti-Humean: dispositional properties – whether they be those of a system's parts or of a system as a whole – are taken to be powers that are irreducibly modal and teleological.

Another important clarification, to avoid misunderstanding, is that a powers-based theory of goal-directedness, and the goal-contribution account of function that goes naturally with it, are perfectly consistent with the fact that some branches of science employ those 'backward-looking' functional explanations that we discussed earlier in Section 2. As we saw, such explanations are pervasive in evolutionary biology. Within our powers-based understanding of goal-directedness, the idea would be that the properties described in such explanations are naturally selected powers, which are essentially directed towards certain goals. In the case of biology, naturally selected biological functions will typically be powers of traits which previously contributed to the goal of the survival of the genes that generate them. These goal-directed powers will then, along with evolutionary theory and environmental facts regarding habitat, explain the prevalence of present tokens of the relevant traits. In short, the powers-based account of teleology fully accommodates naturally selected functions, but as a general metaphysical analysis it also makes room for non-selected functions and goals in other domains.

One way of resisting the powers-based conception of goal-directed systems is to argue that it is in many cases inappropriate to posit power properties at non-fundamental, macro levels (see, e.g., Bird 2016). For example, according to some 'sparse' conceptions of properties, it is only fundamental entities like particles and waves which instantiate genuine properties (see e.g., Armstrong 1978; Heil 2021; Lewis 1983). If we accept this view, then we would have to say that complex entities, such as systems, do not *really* have power properties, even though it is perhaps useful to speak as if they do.

Elsewhere, I have questioned this overly sparse conception of properties (2022a, 2022b). This is not because I accept that all predicates correspond to properties. In my view, we should accept the existence of higher-level, multiply realizable properties only if they can perform certain theoretical and explanatory work. Properties in the special sciences provide salient examples. Many special scientists such as chemists, biologists, and psychologists would accept (or suspect) that the properties they posit depend existentially in some way on the lower-level properties of physics. Nonetheless, the special sciences, and the properties they posit, have an explanatory power that would be lost if we could speak only of the properties of physics. For example, if we want to know what the function of the human thymus is, an explanation involving movements of electrons and protons is unlikely to be illuminating. This is mainly why the special sciences exist in addition to physics and have had so much predictive and explanatory success. Accordingly, recent powers theorists have been more sympathetic to the idea that there are higher-level, macro powers. For example, Vetter (2018) and Mumford (2021) have defended certain kinds of macro powers against criticisms from Bird (2016),[44] while Kimpton-Nye (2022) has argued that if we accept a certain grounding-based conception of powers, a commitment to higher-level powers plausibly follows.

Notice that a more liberal conception of power properties need not commit us to thinking that higher-level properties are strongly emergent or wholly autonomous. Importantly, powers and other properties can do important explanatory work even if they are metaphysically *grounded* in properties that are more fundamental (see Guo and Tugby 2023).[45] Although there is more one could say about these issues, I shall assume in what follows that we should countenance powers at different levels of complexity.[46] Even if higher-level powers are

---

[44] Note, though, that even Bird accepts there are evolved macro powers (2018), and Vetter (2018) argues that Bird's reasons for accepting evolved powers apply equally to powers of artefacts.
[45] This is not to say that higher-level powers are *reducible* to lower-level ones. See Audi 2012 and Tugby 2022a: Ch 6.2 on why grounding does not entail reduction. My view on higher-level powers has affinities with what Vetter calls 'explanatory dispositionalism' (2018: 294; 2021).
[46] For more detailed defences of macro powers, see Kimpton-Nye 2022, Mumford 2021, Tugby 2022a, 2022b, and Guo and Tugby 2023.

metaphysically grounded in lower-level ones, it would be a disservice to the higher-level powers (and to the special sciences generally) to regard them as somehow second-rate. Moreover, if we are in the business of explaining the goal-directedness of directively organized systems, it seems inevitable that we must appeal to higher-level properties of one sort or another. As we saw in Section 3, cybernetic theorists appeal to properties such as plasticity, persistence, or representation, which are unlikely to be involved in fundamental physical theory. So, in the current context, allowing higher-level properties to do explanatory work ought to be relatively uncontroversial.

Let us round of this subsection by highlighting a key benefit of a powers-based account of goal-directedness, which is that it can accommodate some intuitive but subtly different notions of goal-directedness. Here I mention two relevant distinctions; whether there are others is a question I leave open. First, there is arguably an important distinction to be drawn between what we might call 'latent' and 'active' goal-directedness.[47] For example, when a heat-seeking missile is dormant in its holding bay, it is not actively seeking a target even though it is inherently directed towards the destruction of some target. This is goal-directedness in the latent sense. On the other hand, once a missile has been launched, its goal-directedness is no longer latent, for it is now actively seeking out a target. The powers theory can straightforwardly accommodate this intuitive distinction because it is widely acknowledged that powers can exist in a latent state, independently of their activation. Even if a homing missile is not doing anything, its goal-directedness is nonetheless grounded by the relevant latent power.

Second, another distinction concerns what we might call 'immediate' versus 'iterated' goals. For example, we might think that, in some sense, a doctor has both a complex power to heal people and also a complex power to murder people. This might puzzlingly suggest that the doctor has contrary goals, which could create difficulties for practical deliberation. However, we can avoid any puzzles here by distinguishing immediate and iterated powers (see e.g., Vetter 2015: Ch. 4). Assuming that the doctor is benevolent, the *immediate* overriding power at work will be the power to heal. The doctor may well have the physiology required to commit a complex murderous act, but she will not have the power to murder in any immediate sense given her benevolent nature. Nonetheless, this is consistent with the doctor having an iterated power *to acquire* the immediate power to murder. For instance, if the doctor were to undergo radical psychological changes, perhaps she would then gain an operative power to murder. Thus, if goal-directedness is grounded in powers, it is not

---

[47] Here I am grateful for discussions about the examples with Simona Aimar and Toby Friend.

difficult to accommodate this intuitive distinction between immediate and iterated goal-directedness, where the latter concerns possibilities that are much more remote.[48]

4.5 Another Look at the Problem of Goal-Failure: A Powerful Solution

With a powers-based account of goal-directedness now on the table, the important question is whether this account can provide a solution to the problems of goal failure and underdetermination that we discussed at length in Section 3.4. The problem with the reductive cybernetic analysis of goal-directedness, recall, was that it did not provide the resources to make sense of goal-directedness in cases where the goal was not achieved. However, if we accept that goal-directed systems have teleological dispositions in Kroll's sense, then a metaphysically robust notion of end-directedness is built in from the start. And as the consequent of Kroll's (CDC) in Section 4.3 implies, like all other dispositions, there will be no guarantee that the dispositions of directively organized systems always manifest their goals when activated, because certain interferences might prevent those manifestations. If we accept a powers-based account of goal-directedness, goal failure is entirely to be expected in many cases. But even if the relevant goals are not achieved, the dispositions of directively organized systems may nonetheless be responsible for the behaviours that are trying to bring about those goals.

More precisely, if we apply the Kroll-type teleological analysis of dispositions to the powers of directively organized systems, we may say that a directively organized system's powers are such that, when activated, they either reach their goals immediately or initiate a goal-directed process which leads to the goal – unless the process is interrupted. In the case of systems that are *highly* plastic and persistent, there might not be many possible obstacles that could frustrate the goal-directed process. However, failure is always a possibility for any power. Indeed, in some cases a goal-directed system might consistently miss its target, as in Ehring's example (1984a: 502) of a bird that regularly fails to kill its prey due to interfering factors. Such failures are consistent with the bird having the *power* to kill prey, providing that it has, say, the right anatomical and psychological features. This may be so even if the interfering factor is another internal feature of the bird (see e.g., Tugby 2016 for a discussion of dispositions that are subject to so-called intrinsic finks and antidotes).

---

[48] Relatedly, if we also accept a goal-contribution theory of functions, we will have the resources to distinguish between immediate, operative functions and iterated functions. An iterated function would be one that could arise if a certain (immediate) goal were acquired by the relevant system. Similarly, the latent/active distinction can also be applied to functions. A function will be latent if the relevant goal-directed system has not yet been activated.

It is clear, then, that a powers-based theory of goal-directedness can accommodate failure cases, but recall that the problem of goal failure also led to the problem of underdetermination. If the goal of a system is not achieved, then what is it that determines what the goal of the system was meant to be? A powers-based account of goal-directedness delivers a metaphysical answer. The goal is whatever manifestation it is that the system's characteristic (plastic and persistent) powers are directed towards.[49] These powers, in turn, ground the sorts of counterfactuals discussed in Section 3.5. Even if, for whatever reason, a goal fails to be achieved, it can still be the case that the goal *would* have been achieved in certain hypothetical scenarios. Our complaint in Section 3.5 was that it remained unclear what it is in the world that determines the truth of such counterfactuals. But with realist powers in play, we have at least one way of meeting this challenge. For if (CDC) or something like it is correct, then powers have counterfactual implications even if they cannot themselves be analysed in counterfactual terms (see also Mumford 1998: Ch. 4).

To conclude, if a metaphysics of directed powers is accepted, the prospects look good for a solution to the problem of how there can be goal-directedness in cases where the relevant goals fail to occur. And because powers can be intrinsic to their possessors, this view can accommodate the intrinsicality of goal-directedness, which means that it is not committed to the design-based theories of teleology discussed earlier (e.g., Section 3.4), on which the goals of systems are determined by the intentions of a designer such as a human agent or God. Nonetheless, this is not quite the end of the matter. There is still more to be said about the details of a powers-based solution to the problem of goal failure. Despite what has been said so far, some critics of the powers theory have been puzzled about how any intrinsic dispositional state can be directed towards a manifestation that does not (or may never) occur. As we shall see now see, in the powers literature some specific strategies have been explored to deal with this puzzle, one of which entails a commitment to a Platonic conception of properties.

4.6 A Platonic Twist?

I suspect that critics of a powers-based account of goal-directedness will insist that it remains mysterious how *exactly* an intrinsic power, or a process that it initiates, can be directed towards an end or goal that may never occur. Even if

---

[49] A referee points out that even if the relevant goals do not *currently* exist, perhaps it could be maintained in the biological cases that such goals *did* exist when the relevant function was selected. And hence, we can solve the underdetermination problem in those cases by appealing to past events rather than uninstantiated end-states. This is an interesting proposal and I welcome further work on it. For current purposes, I will just note that this approach would not provide a fully general solution to the problem, on the plausible assumption that not all functions are naturally selected.

the end-directedness is primitive, surely we ought nonetheless try to make it intelligible how there can be directedness *of any kind* towards entities (in this case power manifestations) that do not occur. Even if a reductive analysis of end-directedness is out of the question, further explication is surely needed if we are to understand how dispositions can 'point towards' their non-occurring manifestations; or so the objection goes. Notice, though, that it seems unlikely that work in philosophy of science can shed much light on the nature of end-directedness, which is fundamentally a metaphysical issue. There are philosophers of science who have seen the value of appealing to dispositional concepts in accounts of teleological explanation (e.g., Bigelow and Pargetter 1987; Walsh 2008: 119). As far as I can tell, most of these philosophers think that as soon as we appeal to dispositional concepts, puzzles about the forward-looking nature of many teleological explanations automatically go away. However, some critics might remain unconvinced.

Earlier, we saw how the powers-based solution to the problem of goal failure relies on the fact that powers in general can be directed towards non-existent manifestations. However, it is not entirely clear that, for example, Kroll's analysis sheds much light on this aspect of powers. If one is puzzled about how intrinsic dispositions can be directed towards certain manifestations that do not occur, it probably does not help much to be told that a disposition is a primitive, end-directed state. Indeed, some commentators have complained that Kroll's primitive notion of teleological directedness is no better understood than the concept of a disposition itself (Manley and Wasserman 2017: 48; see also Marmodoro 2022: Sect. 1.2).

One of the problems here is that many theories of dispositions or powers leave important questions unanswered. For instance, a simple but important question is whether teleological directedness is relational or not. As I have noted in a different context (Tugby 2013), dispositional directedness certainly looks like a relation: it is a case of an entity being orientated towards something else. If it is a relation, then we could partly shed light on end-directedness by clarifying what are the relata. However, if the directedness of a power or process can occur in the absence of a physical instantiation of the manifestation or goal, it is far from obvious what the relata can be (Armstrong 1997). As Armstrong puts it, it seems that a disposed object 'points to a thing that does not exist' (1997: 79). Prima facie, then, realists about powers are committed to the idea that unmanifested dispositions are directed towards non-existent 'Meinongian' entities. So-called Meinongian entities are certainly taken by many to be obscure, since they do not exist and yet they can somehow be quantified over.[50] If, on the other

---

[50] For a modern discussion of the Meinongian idea, see Azzouni 2004.

hand, directedness is not a genuine relation, how is it that a *non-relational* state can be orientated towards, and individuated by, a merely potential manifestation?

On close inspection, Armstrong's challenge stems from the difficulty of reconciling two facts about dispositions regarding their directedness. Let us call the first fact the 'directedness' principle and the second fact the 'independence' principle:[51]

Directedness Principle: By their very nature, dispositions are directed towards their manifestations.

Independence Principle: An object can instantiate a disposition even if that disposition is never manifested.

Despite their independent plausibility, the conjunction of these two principles generates the puzzle that we have already alluded to. If a disposition can be instantiated in the absence of its manifestation (per the independence principle), how can that disposition at the same time be directed towards the manifestation, as the directedness principle requires? Is there a way to reconcile these two principles while avoiding the Meinongian metaphysics to which Armstrong refers?

According to the Platonic solution that I favour, there is a type–token ambiguity in the principles as stated. When we say that a forever-dry cube of salt has the unmanifested disposition of water-solubility, what we surely mean is that this particular cube of salt has not been engaged in any concrete, token dissolving episodes. The independence principle, then, is best understood in terms of independence *of token manifestations*. In contrast, I take it that the directedness principle concerns manifestations considered as *types*. That is, the directedness principle is a more general principle about the connection of individuation between dispositions and manifestations *qua types* of properties. To use a familiar example, the disposition of fragility is individuated, at least in part, as that disposition which is directed towards the manifestation of breakage. But here we are clearly talking about breakage as a *type* of manifestation, rather than a specific token of breakage. In itself, something's being fragile is compatible with different concrete breakages at different points in space and time; and so, the directedness principle abstracts away from specific token manifestations. Oderberg puts the point well when he describes the directedness of powers as *specific indifference* (2017: Sect. 3). Dispositions are for specific *types* of manifestation but are indifferent regarding the when or where of their *token* manifestations.

---

[51] This tension is also explored in Giannini and Tugby (2020).

How, then, do these clarifications help with our puzzle about non-existent manifestations? The answer lies in adopting a robust Platonic realism about types and viewing dispositional directedness as a second-order affair, holding between *types* rather than tokens. Let us imagine a scenario in which no soluble object ever manifests its solubility in any token dissolving events. Does this mean that we require Meinongian manifestations in order to secure the directedness (and hence identity) of the solubility property? The answer is 'no'. According to a Meinongian approach, such manifestations do not exist even though we can (somehow) quantify over them. However, an alternative option is to accept that the solubility property is directed towards a manifestation *type* (in this case, the property type of dissolving) that exists in the full-blooded, *ontologically committing* sense of the term. This solution commits us to a realist Platonic metaphysics, which has a long history in philosophy. According to that metaphysics, such property types – which include the uninstantiated ones – are actual abstract entities which do not ontologically depend on their concrete instantiations.[52]

It must be acknowledged that Aristotelians are also realists about property types (e.g., Lowe 2006). However, Aristotelians insist that property types must be instantiated *at least once* in order to exist. This means, unfortunately, that Aristotelian realism does not provide enough metaphysical resources to deal with the solubility example described, given that the manifestation property is not instantiated at all. It seems the only option for Aristotelians in such cases is to deny that the relevant disposition exists. This is a significant bullet to bite and a neglected drawback of many Aristotelian accounts of dispositions (see Tugby 2013 for further discussion).

I would urge, therefore, that Platonism is to be preferred when it comes to shedding light on the end-directedness of dispositions. If we apply the Platonic version of dispositional realism to goal-directed systems, then goal-directedness is what we get when systems have characteristic plastic and persistent powers that are essentially directed towards their (Platonic) manifestation-types. Importantly, these types exist regardless of whether they are ever instantiated.[53] This, in turn, allows us to make metaphysical sense of goal-directedness in cases where the relevant goals fail to be achieved.

Despite the power of this argument (no pun intended), many powers theorists appear to be happy with the rival Aristotelian theory of properties. Oderberg

---

[52] Interestingly, comments from McKitrick suggest that Kroll himself might be implicitly committed to uninstantiated types, given that his analysis of dispositions quantifies over stimulus and manifestation types (see McKitrick 2017: 43 for discussion).

[53] Note that although I regard manifestations as property-types, the Platonic theory can equally be expressed in terms of event-types. For current purposes, nothing important hangs on this detail.

(2017: 2403), for example, has argued that Aristotelianism offers an adequate framework for unmanifested powers and that Platonism does not provide significant theoretical gains.[54] Others, meanwhile, have accounted for unmanifested manifestations in terms of either unrealized possibilia or mere logical existents (see, e.g., Bird 2006; Giannini 2021; McKitrick 2018: Ch. 3).[55] I shall leave the issue at this point and allow readers to judge for themselves. What we have at least shown is that Platonism provides one possible way of shedding light on the directedness of powers. In Section 4.8, we shall also see how Platonism proves to be helpful when it comes to accommodating certain cases of goal-directedness involving physically impossible goals.

In the meantime, let us broach an important question regarding the extent of goal-directed teleology.

4.7 Further Work: The Extent of Teleology

A noteworthy implication of the powers theory is that goal-directedness is not the only teleological phenomenon. If Kroll's theory is along the right lines, then *all* dispositional states are essentially end-directed, and this applies as much to the powers of simples, like electrons, as to those of complex systems. On this view, the character of any concrete causal interaction is ultimately explained by the end-directedness of the powers involved.[56] And if many powers are essential to their possessors, we will have a robust metaphysical foundation for normativity in the case of systemic goal-directed processes. For if the relevant powers are what characterize the systems in question, there is an important sense in which those systems *should* exhibit the relevant goal-directed behaviour if they are functioning normally. For further discussion of the connection between dispositions, essence, and natural normativity, see for example, Lowe (1980, 1982, 1987), Korsgaard (2009: Ch. 2), and Oderberg (2010, 2020).

Now, although all dispositions are end-directed, it does not follow that all dispositions are *goal-directed*. Rather, our previous discussions suggest that goal-directedness is what we get when dispositional end-directed states are instantiated at a certain level of systemic complexity. However, this leaves us with interesting and difficult questions about where to draw the line. What exactly distinguishes goal-directed dispositions from the more basic end-

---

[54] But see Tugby 2022a: 64 for a response. For discussion of further theoretical benefits of Platonism in the context of science, see e.g., Brown 1991, 1994, Berman 2020: Ch. 6, and Tugby 2022a.

[55] As Giannini defines them (2021: 2684), mere logical existents are non-essentially non-located entities, i.e., neither purely abstract nor purely concrete.

[56] This idea arguably rejuvenates the Aristotelian notion of final causation. I do not have the space to explore this connection in detail, but interested readers should see Oderberg 2017 on the connection between power realism and final causation. See also Feser 2014 and 2019: Ch. 6

directed dispositions that we would not regard as being goal-directed? This is a complex question, and I am not able to fully settle it here. Nonetheless, aspects of the cybernetic theory of goal-directedness might prove helpful. According to that theory, what many (if not all) goal-directed systems have in common is that they can display plastic and persistent behaviour. In dispositional terms, the idea is that systemic, goal-directed dispositions have a modal flexibility and adaptability that more primitive end-directed dispositional states lack. Unlike a homing missile, if a marble is rolling towards you and someone pushes it off course, it will not adjust its direction of travel in such a way as to maintain its original course. Although a marble's state of sphericality is dispositionally directed towards rolling, one could not plausibly regard this as a goal-directed state.

There is a complication, though, which is that plasticity and persistence arguably come in degrees (Section 3.2). In Nagel's view, a system's degree of directive organization depends on how many different internal and external variations it is able to compensate for when pursuing its goal (Nagel 1961: 417). As Garson notes (2016: 24), a similar idea can be found in the work of Braithwaite, whose analysis concerns the range or 'variancy' of environmental field-conditions in which a system is able to achieve a given type of goal (1953: 330). According to Braithwaite, a system is plastic merely if it is able to achieve a certain goal in more than one alternative set of field-conditions (1953: 331).

If plasticity and persistence come in degrees, one might well wonder where the cut-off points lie. To what degree must a system's overall powers be plastic and persistent in order for it to count as genuinely goal-directed? The goal-contribution account of functions allows for a variety of stances. Braithwaite for one sets the bar quite low, allowing that a goal-directed system might be plastic to a rather minimal degree. Nagel appears to be more ambivalent, suggesting that there might not be a clear cut-off point (1961: 419). If Nagel is right, this might explain why there are so many disagreements about which systems can and cannot be ascribed functions (see Oderberg 2008; McShea 2012; Tugby 2020; Babcock 2023 for discussion). For my part, I am inclined to take a liberal stance, and accept that there can be goal-directedness even if the degree of plasticity and persistence is relatively low. However, I will have to leave this as a matter for further in-house discussion between goal-contribution theorists.

4.8 Further Work: Can There Be Goal-Directedness without Goal-Directed Powers?

So far, we have argued that a powers-based account of goal-directedness can provide a plausible account of goal-directedness in many cases. In doing so, we utilized some insights from the philosophy of science literature regarding cybernetic machines and the notions of plasticity and persistence. However, whether the powers-based account can provide a fully general analysis of *all* kinds of goal-directedness is a further question.

The most obvious sticking points for a fully general powers-based account of goal-directedness are cases allegedly involving goal-directedness in which the relevant goals are in some sense impossible. Ehring has already presented cases of impossible goals as counterexamples to the Nagelian analysis of goal-directedness, such as the example of Smith who apparently has the goal of accelerating a spaceship beyond the speed of light (1984a: 500).[57] Importantly, it looks as though these same cases might create an equally serious problem for a generalized powers-based account of goal-directedness. The worry is simply that in such cases there is genuine goal-directedness but no powers to bring about the relevant goals. This worry is driven by the thought that it is implausible to posit powers for impossible goals. One will find this especially implausible if, like Vetter (2015), one thinks that there is a tight connection between powers and possibility.

The issue of powers with impossible goals is complex and we shall not be able to do it full justice here. However, we can at least set out some options for powers theorists and lay foundations for further discussion of the issue. In the process, I shall explain how a Platonic theory of powers might accommodate powers with impossible goals in many cases.

The question of whether there are powers for impossible manifestations has recently attracted attention, and there are philosophers who fall on both sides of the fence. For example, Jenkins and Nolan (2012) offer theoretical reasons for positing a wide range of dispositions with impossible manifestations. Vetter (2015), in contrast, draws a close metaphysical connection between powers and possibility, and therefore denies that there are powers for impossible manifestations.

Clearly, if we were to agree that there are no powers for impossible manifestations, while accepting cases of goal-directedness towards impossible goals, we would have to deny that all cases of goal-directedness are grounded in powers. In order to accommodate this idea, one could take inspiration from Scheffler (1959) and try to draw a distinction between cases of goal-directedness involving an agent's intentional action, and those that do not. Perhaps what the powers-based theory offers is a metaphysical account of goal-directedness in cases *not* involving intentional action. Cases of intentional action, it might be urged, involve a distinct, *sui generis* form of goal-directedness. Importantly, we could accommodate impossible goals in such cases as long as an agent can have intentions with impossible intentional objects. According to that strategy, then,

---

[57] There is a related debate about how (if at all) we can make sense of counterfactuals with nomically or metaphysically impossible antecedents or consequents (e.g., Tan 2017, 2019, and Kimpton-Nye 2020).

goal-directed behaviour towards impossible goals does occur, but only when the behaviour is driven by a (misguided) intention. The disadvantage of that approach, however, is that it leaves us with a somewhat disunified analysis of goal-directedness.

What other options are there? Elsewhere, I have provided independent arguments for the existence of Platonic types whose instantiations are physically/nomically impossible in worlds like ours (see e.g., Tugby 2015, 2022a: Ch. 10, and Giannini and Tugby 2020). In line with terminology employed by Vetter (2015: 69), I call such properties 'super-alien' properties, and possible examples include the property of being frictionless (Giannini and Tugby 2020: 133–134). Given that scientists often engage in modal reasoning about super-alien properties, and given that scientists appear to have non-trivial true beliefs about them, Platonism offers a plausible way to ground such truths in reality. Quite simply, such truths are grounded in the natures of the uninstantiated super-alien properties. If this argument succeeds, then we should count super-alien properties among the abstract types that exist. This, in turn, makes room for the idea that a property instantiated in our world might be dispositionally directed towards a super-alien manifestation type. For instance, following Jenkins and Nolan (2012: 746), we might accept that a car has the power to move in a certain way on a frictionless plane, even though our physical world could never give rise to the manifestation of such a power. Similarly, in Ehring's example, Platonism opens up the possibility that Smith's behaviour is directed towards the super-alien, physically impossible property of travelling faster than the speed of light.

On the other hand, there will still be *some* limitations regarding how many super-alien Platonic properties there are. For example, it is difficult to see how properties could exist whose instantiations are metaphysically or logically impossible (Tugby 2015: 35). If there are no such properties, then a Platonic approach to powers would still not accommodate cases of goal-directed behaviour towards outcomes which are impossible in this stronger logical sense. An example of such a case would be behaviour that is allegedly directed towards proving a mathematical conjecture that is necessarily false. Whether there can be genuine cases of goal-directedness like this is an interesting question that I leave open for future research in this relatively new area of the debate.[58] If there could be cases of goal-directed behaviour towards logically impossible goals, we might well have to accept that those cases involve a *sui generis* form of goal-directedness that does not involve a corresponding power.

---

[58] One perspective on such examples, which I am unable to discuss here, is to treat logical impossibilities along fictionalist lines. For discussion of fictionalism, see Kimpton-Nye 2020.

4.9 Summing Up

The central aims of this section were to show that the powers metaphysics brings a new source of teleology to the world and that it provides better resources for explaining goal-directedness than the reductive cybernetic accounts examined in Section 3. Unlike those accounts, the powers-based theory takes teleological directedness to be a fundamental, irreducible feature of the world. This view thus rejects the reductive aims of goal theorists like Nagel, who tried to explain teleological truths in wholly non-teleological terms. In Section 3, we saw that the reductive project faces serious problems, in particular the problem of goal failure and underdetermination. One of the key lessons to draw from these problems, I believe, is that the prospects for a reductive analysis of teleology are dim. As soon as we accept that a system could be goal directed without ever achieving that goal, we are pushed towards accepting that the system is in a genuine teleological, goal-directed state. In line with the goal-contribution theory of functions, functions can then be ascribed to those parts of a system which have dispositions to make causal contributions to the system's goal-directed behaviour. Suitably modified, contemporary realism about powers provides one way of fleshing out this metaphysical proposal. Importantly, this powers-based, goal-theoretic account is consistent with the sorts of 'backward-looking' selectionist functional explanations that we find in evolutionary biology. The theory can also accommodate functions in those domains of science where many functions are not selected for. However, whether *all* possible cases of functions and goal-directedness can be appropriately grounded in powers is an important question that I leave for future research.

References

Adams, F. R. (1979). A goal-state theory of function attributions. *Canadian Journal of Philosophy* 9: 492–518.

Alvarado, J. T. and Tugby, M. (2021). A problem for natural-kind essentialism and formal causes. In L. Jansen and P. Sandstad (Eds.), *Neo-Aristotelian Perspectives on Formal Causation*, pp. 201–221. Abingdon: Routledge.

Amundson, R. and Lauder, G. V. (1994). Function without purpose: The uses of causal role function in evolutionary biology. *Biology and Philosophy* 9: 443–469.

Anjum, R. L. and Mumford, S. (2018). *What Tends to Be: The Philosophy of Dispositional Modality*. Abingdon: Routledge.

Anscombe, G. E. M. (1968). The intentionality of sensation: A grammatical feature. In R. J. Butler (Ed.), *Analytical Philosophy, Second Series*, pp. 158–180. Oxford: Blackwell.

Armstrong, D. M. (1978). *Universals and Scientific Realism*, Vols. I and II. Cambridge: Cambridge University Press.

Armstrong, D. M. (1997). *A World of States of Affairs*. Cambridge: Cambridge University Press.

Armstrong, D. M. (2004). *Truth and Truthmakers*. Cambridge: Cambridge University Press.

Audi, P. (2012). A clarification and defense of the notion of grounding. In F. Correia and B. Schnieder (Eds.), *Metaphysical Grounding: Understanding the Structure of Reality*, pp. 101–121. Cambridge: Cambridge University Press.

Austin, C. J. (2017). A biologically informed hylomorphism. In W. M. R. Simpson, R. C. Koons, and N. J. Teh (Eds.), *Neo-Aristotelian Perspectives on Contemporary Science*, pp. 185–209. Abingdon: Routledge.

Austin, C. J. and Marmodoro, A. (2017). Structural powers and the homeodynamic unity of organisms. In W. M. R. Simpson, R. C. Koons, and N. J. Teh (Eds.), *Neo-Aristotelian Perspectives on Contemporary Science*, pp. 169–184. Abingdon: Routledge.

Azzouni, J. (2004). *Deflating Existential Consequence: A Case for Nominalism*. New York: Oxford University Press.

Babcock, G. (2023). Teleology and function in non-living nature. *Synthese* 201, 112: 1–20.

Barandiaran, X. and Moreno, A. (2008). Adaptivity: From metabolism to behavior. *Adaptive Behavior* 16: 325–344.

Bauer, W. A. (2022). *Causal Powers and the Intentionality Continuum*. Cambridge: Cambridge University Press.

Beckner, M. (1959). *The Biological Way of Thought*. New York: Columbia University Press.

Bedau, M. (1992a). Where's the good in teleology? *Philosophy and Phenomenological Research* 52: 781–806.

Bedau, M. (1992b). Goal-directed systems and the good. *Monist* 75: 34–51.

Bellazzi, F. (2022). The emergence of the postgenomic gene. *European Journal for the Philosophy of Science* 12: 1–21.

Bellazzi, F. (forthcoming). Biochemical functions. *British Journal for the Philosophy of Science*.

Berman, S. (2020). *Platonism and the Objects of Science*. London: Bloomsbury.

Bigelow, J. and Pargetter, R. (1987). Functions. *Journal of Philosophy* 84: 181–196.

Bird, A. (2006). Potency and modality. *Synthese* 149: 491–508.

Bird, A. (2007). *Nature's Metaphysics: Laws and Properties*. Oxford: Oxford University Press.

Bird, A. (2016). Overpowering: How the powers ontology has overreached itself. *Mind* 125: 341–383.

Bird, A. (2018). Fundamental powers, evolved powers, and mental powers. *Proceedings of the Aristotelian Society Supplementary Volume* 92: 247–275.

Boorse, C. (1976). Wright on functions. *Philosophical Review* 85: 70–86.

Boorse, C. (1977). Health as a theoretical concept. *Philosophy of Science* 44: 542–573.

Boorse, C. (2002). A rebuttal on functions. In A. Ariew, R. Cummins, and M. Perlman (Eds.), *Functions: New Essays in the Philosophy of Psychology and Biology*, pp. 63–112. Oxford: Oxford University Press.

Bouchard, F. (2013). How ecosystem evolution strengthens the case for function pluralism. In P. Huneman (Ed.), *Function: Selection and Mechanisms*, pp. 83–95. Dordrecht: Springer.

Bourrat, P. (2021). Function, persistence, and selection: Generalizing the selected-effect account of function adequately. *Studies in History and Philosophy of Science* 90: 61–67.

Braithwaite, R. B. (1953). *Scientific Explanation*. Cambridge: Cambridge University Press.

Brandon, R. N. (2013). A general case for function pluralism. In P. Huneman (Ed.), *Function: Selection and Mechanisms*, pp. 97–104. Dordrecht: Springer.

Brentano, F. (2015/1874). *Psychology from an Empirical Standpoint*. Abingdon: Routledge.

Brown, J. R. (1991). *The Laboratory of the Mind: Thought Experiments in the Natural Sciences*. Abingdon: Routledge.

Brown, J. R. (1994). *Smoke and Mirrors: How Science Reflects Reality.* Abingdon: Routledge.

Canfield, J. (1964). Teleological explanations in biology. *British Journal for the Philosophy of Science* 14: 285–295.

Cartwright, N. (1986). Two kinds of teleological explanation. In A. Donagan, A. N. Perovich, Jr., and M. I. Wedin (Eds.), *Human Nature and Natural Knowledge*, pp. 201–210. Vol. 89 of *Boston Studies in the Philosophy of Science*. Dordrecht: D. Reidel.

Cartwright, N. (1992). Aristotelian natures and the modern scientific method. In J. Earman (Ed.), *Inference, Explanation, and Other Frustrations*, pp. 44–71. California: University of California Press.

Cartwright, N. (1999). *The Dappled World*. Cambridge: Cambridge University Press.

Cartwright, N. (2019). *Nature, the Artful Modeler: Lectures on Laws, Science, How Nature Arranges the World and How We Can Arrange It Better.* Chicago, IL: Open Court.

Cartwright, N. and Pemberton, J. (2013). Aristotelian powers: Without them, what would modern science do? In J. Greco and R. Groff (Eds.), *Powers and Capacities in Philosophy: The New Aristotelianism*, pp. 93–112. New York: Routledge.

Christensen, W. D. and Bickhard, M. H. (2002). The process dynamics of normative function. *The Monist* 85: 3–28.

Corry, R. (2019). *Power and Influence: The Metaphysics of Reductive Explanation*. Oxford: Oxford University Press.

Coyne, L. (2016). Phenomenology and teleology: Hans Jonas's philosophy of life. *Environmental Values* 26: 297–315.

Crane, T. (2008). Reply to Nes. *Analysis* 68: 215–218.

Craver, C. (2013). Functions and mechanisms: A perspectivalist view. In P. Huneman (Ed.), *Function: Selection and Mechanisms*, pp. 133–158. Dordrecht: Springer.

Cummins, R. (1975). Functional analysis. *Journal of Philosophy* 72: 741–765.

Doolittle, W. F. (2013). Is junk DNA bunk? A critique of ENCODE. *Proceedings of the National Academy of Sciences USA* 110: 5294–5300.

Ehring, D. (1984a). The system-property theory of goal-directed processes. *Philosophy of the Social Sciences* 14: 497–504.

Ehring, D. (1984b). Negative feedback and goals. *Nature and System* 6: 217–220.

Ellis, B. (2001). *Scientific Essentialism*. Cambridge: Cambridge University Press.

Ellis, B. and Lierse, C. (1994). Dispositional essentialism. *Australasian Journal of Philosophy* 72: 27–44.

Faber, R. J. (1986). *Clockwork Garden: On the Mechanistic Reduction of Living Things*. Amherst: University of Massachusetts Press.

Feser, E. (2009). *Aquinas*. Oxford: Oneworld.

Feser, E. (2014). *Scholastic Metaphysics*. Heusenstamm: Editiones Scholasticae.

Feser, E. (2019). *Aristotle's Revenge: The Metaphysical Foundations of Physical and Biological Science*. Neunkirchen: Editiones Scholasticae.

Forber, P. (2020). Contemporary teleology. In J. K. McDonough (Ed.), *Teleology: A History*, pp. 255–278. Oxford: Oxford University Press.

Franklin, A. and Knox, E. (2018). Emergence without limits: The case of phonons. *Studies in History and Philosophy of Science Part B: Studies in History and Philosophy of Modern Physics* 64: 68–78.

Friend, T. and Kimpton-Nye, S. (2023). *Dispositions and Powers*. Cambridge: Cambridge University Press.

Gambarotto, A. (2020). Teleology, life and cognition: Reconsidering Jonas's legacy for a theory of the organism. In A. Altobrando and P. Biasetti (Eds.), *Natural Born Monads: On the Metaphysics of Organisms and Human Individuals*, pp. 243–264. Berlin: De Gruyter.

Garson, J. (2016). *A Critical Overview of Biological Functions*. Dordrecht: Springer.

Garson, J. (2017). A generalized selected effects theory of function. *Philosophy of Science* 84: 523–543.

Garson, J. (2018). How to be a function pluralist. *British Journal for the Philosophy of Science* 69: 1101–1122.

Garson, J. (2019a). *What Biological Functions Are and Why They Matter*. Cambridge: Cambridge University Press.

Garson, J. (2019b). There are no ahistorical theories of function. *Philosophy of Science* 86: 1146–1156.

Garson, J. and Piccinini, G. (2014). Functions must be performed at appropriate rates in appropriate situations. *British Journal for the Philosophy of Science* 65: 1–20.

Germain, P.-L., Ratti, E., and Boem, F. (2014). Junk or functional DNA? ENCODE and the function controversy. *Biology and Philosophy* 29: 807–831.

Giannini, G. (2021). New powers for dispositionalism. *Synthese* 199: 2671–2700.

Giannini, G. and Mumford, S. (2021). Formal causes for powers theorists. In L. Jansen and P. Sandstad (Eds.), *Neo-Aristotelian Perspectives on Formal Causation*, pp. 87–105. Abingdon: Routledge.

Giannini, G. and Tugby, M. (2020). Potentiality: Actualism minus Naturalism equals Platonism. *Philosophical Inquiries* 8: 117–140.

Giannotti, J. (2021). The identity theory of powers revised. *Erkenntnis* 86: 603–621.

Giroux, E. (2015). Epidemiology and the bio-statistical theory of disease: A challenging perspective. *Theoretical Medicine and Bioethics* 36: 175–195.

Godfrey-Smith, P. (1993). Functions: Consensus without unity. *Pacific Philosophical Quarterly* 74: 196–208.

Goff, P. (2017). *Consciousness and Fundamental Reality*. Oxford: Oxford University Press.

Gould, S. J. and Vrba, E. S. (1982). Exaptation – A missing term in the science of form. *Paleobiology* 8: 4–15.

Griffiths, P. E. (1993). Functional analysis and proper function. *British Journal for the Philosophy of Science* 44: 409–422.

Griffiths, P. E. and Stotz, K. (2013). *Genetics and Philosophy: An Introduction*. Cambridge: Cambridge University Press.

Guo, X.-Y. and Tugby, M. (2023). Collective powers. In C. J. Austin, A. Marmodoro, and A. Roselli (Eds.), *Powers, Parts and Wholes: Essays on the Mereology of Powers*, pp. 142–166. Abingdon: Routledge.

Hawthorne, J. and Nolan, D. (2006). What would teleological causation be? In J. Hawthorne (Ed.), *Metaphysical Essays*, pp. 265–283. Oxford: Oxford University Press.

Hegel, G. W. F. (2010/1816). *The Science of Logic*. G. di Giovanni (trans.). Cambridge: Cambridge University Press.

Heil, J. (2021). *Appearance in Reality*. Oxford: Oxford University Press.

Hildebrand, T. (2023). *Laws of Nature*. Cambridge: Cambridge University Press.

Hull, D. (1973). *Philosophy of Biological Science*. Engelwood Cliffs, NJ: Prentice-Hall.

Jenkins, C. S. and Nolan, D. (2012). Disposition impossible. *Noûs* 46: 732–753.

Jonas, H. (2001). *The Phenomenon of Life: Toward a Philosophical Biology*. Evanston, IL: Northwestern University Press.

Kant, I. (2007/1790). *Critique of Judgement*. J. C. Meredith (trans.). Oxford: Oxford University Press.

Kertész, G. and Kodaj, D. (2023). In defense of teleological intuitions. *Philosophical Studies* 180: 1421–1437.

Kimpton-Nye, S. (2020). Necessary laws and the problem of counterlegals. *Philosophy of Science* 87: 518–535.

Kimpton-Nye, S. (2021). Reconsidering the dispositional essentialist canon. *Philosophical Studies* 178: 3421–3441.

Kimpton-Nye, S. (2022). Pandispositionalism and the metaphysics of powers. *Synthese* 200, 371: 1–21.

Kistler, M. (2006). *Causation and Laws of Nature*. Abingdon: Routledge.

Kitcher, P. (1993). Function and design. *Midwest Studies in Philosophy* 18: 379–397.

Koons, R. C. and Pruss, A. (2017). Must functionalists be Aristotelians? In J. D. Jacobs (Ed.), *Causal Powers*, 194–204. Oxford: Oxford University Press.

Korsgaard, C. M. (2009). *Self-Constitution: Agency, Identity, and Integrity*. Oxford: Oxford University Press.

Kroll, N. (2017). Teleological dispositions. *Oxford Studies in Metaphysics* 10: 1–37.

Lee, J. G. and McShea, D. W. (2020). Operationalizing goal directedness: An empirical route to advancing a philosophical discussion. *Philosophy, Theory and Practice in Biology* 12, 5: 1–31.

Lewis, D. (1983). New work for a theory of universals. *Australasian Journal of Philosophy* 61: 343–377.

Lewis, D. (1997). Finkish dispositions. *Philosophical Quarterly* 47: 143–158.

Lowe, E. J. (1980). Sortal terms and natural laws: An essay on the ontological status of the laws of nature. *American Philosophical Quarterly* 17: 253–260.

Lowe, E. J. (1982). Laws, dispositions and sortal logic. *American Philosophical Quarterly* 19: 41–50.

Lowe, E. J. (1987). What is the 'problem of induction'? *Philosophy* 62: 325–340.

Lowe, E. J. (2006). *The Four-Category Ontology: A Metaphysical Foundation for Natural Science*. Oxford: Oxford University Press.

Lowe, E. J. (2010). On the individuation of powers. In A. Marmodoro (Ed.), *The Metaphysics of Powers: Their Grounding and Their Manifestations*, pp. 8–26. Abingdon: Routledge.

Makin, S. (2006). *Aristotle: Metaphysics Book Θ*. Oxford: Clarendon Press.

Manier, E. (1971). Functionalism and the negative feedback model in biology. *Boston Studies in the Philosophy of Science* 8: 225–240.

Manley, D. and Wasserman, R. (2017). Dispositions without teleology. *Oxford Studies in Metaphysics* 10: 47–60.

Marmodoro, A. (2022). What's dynamic about causal powers? A black box! In C. J. Austin, A. Marmodoro, and A. Roselli (Eds.), *Powers, Time and Free Will*, pp. 1–15. Cham: Springer.

Martin, C. B. (1993). Power for realists. In J. Bacon, K. Campbell, and L. Reinhardt (Eds.), *Ontology, Causality, and Mind: Essays in Honour of D. M. Armstrong*, pp. 75–86. Cambridge: Cambridge University Press.

Martin, C. B. (1994). Dispositions and conditionals. *The Philosophical Quarterly* 44: 1–8.

Martin, C. B. (2008). *The Mind in Nature*. Oxford: Oxford University Press.

Martin, C. B. and Pfeifer, K. (1986). Intentionality and the non-psychological. *Philosophy and Phenomenological Research* 46: 531–554.

Maturana, H. and Varela, F. (1980). *Autopoiesis and Cognition: The Realization of the Living*. Vol. 42 of *Boston Studies in the Philosophy of Science*. Dordrecht: D. Reidel.

Mayr, E. (1988). The multiple meanings of teleological. In E. Mayr (Ed.), *Towards a New Philosophy of Biology*, pp. 38–66. Cambridge, MA: Harvard University Press.

McDonough, J. K. (Ed.), (2020a). *Teleology: A History*. Oxford: Oxford University Press.

McDonough, J. K. (2020b). Not dead yet: Teleology and the 'scientific revolution'. In J. K. McDonough (Ed.), *Teleology: A History*, pp. 150–179. Oxford: Oxford University Press.

McKitrick, J. (2003). A case for extrinsic dispositions. *Australasian Journal of Philosophy* 81: 155–174.

McKitrick, J. (2017). Indirect directedness. *Oxford Studies in Metaphysics* 10: 38–46.

McKitrick, J. (2018). *Dispositional Pluralism*. Oxford: Oxford University Press.

McLaughlin, P. (2001). *What Functions Explain: Functional Explanation and Self-Reproducing Systems*. Cambridge: Cambridge University Press.

McShea, D. W. (2012). Upper-directed systems: A new approach to teleology in biology. *Biology and Philosophy* 27: 663–684.

Meincke, A. S. (2019). Autopoiesis, biological autonomy and the process view of life. *European Journal for Philosophy of Science* 9, 5: 1–16.

Melander, P. (1997). *Analyzing Functions: An Essay on a Fundamental Notion in Biology*. Stockholm: Almqvist & Wiksell.

Miller, J. F. A. P. (1961). Immunological function of the thymus. *Lancet* 2: 748–749.

Miller, J. F. A. P. (1971). The immunological role of the thymus. In M. Samter (Ed.), *Immunological Diseases* 2nd ed, pp. 84–94. Boston, MA: Little, Brown.

Millikan, R. G. (1984). *Language, Thought, and Other Biological Categories: New Foundations for Realism*. Cambridge, MA: MIT Press.

Millikan, R. G. (1989). In defense of proper functions. *Philosophy of Science* 56: 288–302.

Millikan, R. G. (1993). *White Queen Psychology and Other Essays for Alice*. Cambridge, MA: MIT Press.

Mitchell, S. D. (1993). Dispositions or etiologies? A comment on Bigelow and Pargetter. *The Journal of Philosophy* 90: 249–259.

Molnar, G. (2003). *Powers: A Study in Metaphysics*. S. Mumford (Ed.). Oxford: Oxford University Press.

Mossio, M. and Bich, L. (2017). What makes biological organisation teleological? *Synthese* 194: 1089–1114.

Mossio, M., Saborido, C., and Moreno, A. (2009). An organizational account of biological functions. *British Journal for the Philosophy of Science* 60: 813–841.

Mumford, S. (1998). *Dispositions*. Oxford: Oxford University Press.

Mumford, S. (1999). Intentionality and the physical: A new theory of disposition ascription. *The Philosophical Quarterly* 49: 215–225.

Mumford, S. (2004). *Laws in Nature*. London: Routledge

Mumford, S. (2006). The ungrounded argument. *Synthese* 149: 471–489.

Mumford, S. (2021). Where the real power lies: A reply to Bird. *Mind* 130: 1295–1308.

Nagel, E. (1961). *The Structure of Science*. New York: Harcourt, Brace and World.

Nagel, E. (1979). *Teleology Revisited and Other Essays in the Philosophy and History of Science*. New York: Columbia University Press.

Neander, K. (1983). Abnormal psychobiology. PhD dissertation, La Trobe University.

Neander, K. (1991). Functions as selected effects: The conceptual analyst's defense. *Philosophy of Science* 58: 168–184.

Neander, K. (2002). Why history matters: Four theories of functions. In W. Weingarten and G. Schlosser (Eds.), *Formen der Erklärung in der Biologie*, pp. 91–120. Berlin: VWB-Verlag für Wissenschaft Und Bildung.

Neander, K. (2017). Functional analysis and the species design. *Synthese* 194: 1147–1168.

Nes, A. (2008). Are only mental phenomena intentional? *Analysis* 68: 205–215.

Nissen, L. (1981). Nagel's self-regulation analysis of teleology. *The Philosophical Forum* 12: 128–138.

Nissen, L. (1997). *Teleological Language in the Life Sciences*. Lanham, MD: Rowman and Littlefield.

Oderberg, D. S. (2007). *Real Essentialism*. Abingdon: Routledge.

Oderberg, D. S. (2008). Teleology: Inorganic and organic. In A. M. González (Ed.), *Contemporary Perspectives on Natural Law*, pp. 259–279. Aldershot: Ashgate.

Oderberg, D. S. (2010). The metaphysical foundations of natural law. In H. Zaborowski (Ed.), *Natural Moral Law in Contemporary Society*, pp. 44–75. Washington, DC: Catholic University of America Press.

Oderberg, D. S. (2017). Finality revived: Powers and intentionality. *Synthese* 194: 2387–2425.

Oderberg, D. S. (2020). *The Metaphysics of Good and Evil*. London: Routledge.

Oliver, S. (2013). Aquinas and Aristotle's teleology. *Nova et Vetera* 11: 849–870.

Page, B. (2015). The dispositionalist deity: How God creates laws and why theists should care. *Zygon* 50: 113–137.

Page, B. (2021). Power-ing up neo-Aristotelian natural goodness. *Philosophical Studies* 178: 3755–3775.

Paolini Paoletti, M. (2021a). Functional powers. In L. Jansen and P. Sandstad (Eds.), *Neo-Aristotelian Perspectives on Formal Causation*, pp. 124–148. Abingdon: Routledge.

Paolini Paoletti, M. (2021b). Teleological powers. *Analytic Philosophy* 62: 336–358.

Pittendrigh, C. S. (1958). Adaptation, natural selection and behavior. In A. Roe and G. G. Simpson (Eds.), *Behavior and Evolution*, pp. 390–416. New Haven, CT: Yale University Press.

Place, U. T. (1996). Intentionality as the mark of the dispositional. *Dialectica* 50: 91–120.

Plantinga, A. (1993). *Warrant and Proper Function*. New York: Oxford University Press.

Preston, B. (1998). Why is a wing like a spoon? A pluralist theory of function. *Journal of Philosophy* 95: 215–254.

Prior, E. (1985). What is wrong with etiological accounts of biological function? *Pacific Philosophical Quarterly* 66: 310–328.

Raimondi, A. (2021). Crane and the mark of the mental. *Analysis* 81: 683–693.

Ransome Johnson, M. (2005). *Aristotle on Teleology*. Oxford: Oxford University Press.

Rose, D. and Schaffer, J. (2017). Folk mereology is teleological. *Noûs* 51: 238–270.

Rose, D., Schaffer, J., and Tobia, K. (2020). Folk teleology drives persistence judgments. *Synthese* 197: 5491–5509.

Rosenblueth, A. and Wiener, N. (1950). Purposeful and non-purposeful behavior. *Philosophy of Science* 17: 318–326.

Rosenblueth, A., Wiener, N., and Bigelow, J. (1943). Behavior, purpose and teleology. *Philosophy of Science* 10: 18–24.

Ruse, M. (1971). Functional statements in biology. *Philosophy of Science* 38: 87–95.

Russell, E. S. (1945). *The Directiveness of Organic Activities*. Cambridge: Cambridge University Press.

Sachs, C. (2023). Naturalized teleology: Cybernetics, organization, purpose. *Topoi* 42: 781–779.

Sangiacomo, A. (2015). Teleology and agreement in nature. In A. S. Campos (Ed.), *Spinoza: Basic Concepts*, pp. 87–102. Exeter: Imprint Academic.

Schaffner, K. F. (1993). *Discovery and Explanation in Biology and Medicine.* Chicago, IL: University of Chicago Press.

Scheffler, I. (1959). Thoughts on teleology. *British Journal for the Philosophy of Science* 9: 265–284.

Scheffler, I. (1963). *The Anatomy of Inquiry.* New York: Alfred A. Knopf.

Schelling, F. J. (2000/1800) *System des Transzendentalen Idealismus.* Hamburg: Felix Meiner Verlag.

Schmid, S. (2011). Teleology and the dispositional theory of causation in Thomas Aquinas. *Logical Analysis and History of Philosophy* 14: 21–39.

Schrenk, M. (2016). *Metaphysics of Science: A Systematic and Historical Introduction.* London: Routledge.

Schwartz, J. (1993). Functional explanation and metaphysical individualism. *Philosophy of Science* 60: 278–301.

Smart, B. (2016). *Concepts and Causes in the Philosophy of Disease.* Basingstoke: Palgrave Macmillan.

Sommerhoff, G. (1950). *Analytical Biology.* London: Oxford University Press.

Sorabji, R. (1964). Function. *The Philosophical Quarterly* 14: 289–302.

Spinoza, B. (2018/1677). *Ethics.* M. J. Kisner (Ed.) and M. Silverthorne (trans.). Cambridge: Cambridge University Press.

Stovall, P. (2024). The teleological modal profile and subjunctive background of organic generation and growth. *Synthese* 203, 77: 1–37.

Tahko, T. E. (2020). Where do you get your protein? Or: Biochemical realization. *British Journal for the Philosophy of Science* 71: 799–825.

Tan, P. (2017). Interventions and counternomic reasoning. *Philosophy of Science* 84: 956–969.

Tan, P. (2019). Counterpossible non-vacuity in scientific practice. *The Journal of Philosophy* 116: 32–60.

Taylor, R. (1950a). Comments on a mechanistic conception of purposefulness. *Philosophy of Science* 17: 310–317.

Taylor, R. (1950b). Purposeful and non-purposeful behavior: A rejoinder. *Philosophy of Science* 17: 327–332.

Trestman, M. A. (2012). Implicit and explicit goal-directedness. *Erkenntnis* 77: 207–236.

Tugby, M. (2013). Platonic dispositionalism. *Mind* 122: 451–480.

Tugby, M. (2015). The alien paradox. *Analysis* 75: 28–37.

Tugby, M. (2016). On the reality of intrinsically finkable dispositions. *Philosophia* 44: 623–631.

Tugby, M. (2020). Organic powers. In A. S. Meincke (Ed.), *Dispositionalism: Perspectives from Metaphysics and Philosophy of Science*, pp. 213–238. Cham: Springer.

Tugby, M. (2021). Grounding theories of powers. *Synthese* 198: 11187–11216.

Tugby, M. (2022a). *Putting Properties First: A Platonic Metaphysics for Natural Modality*. Oxford: Oxford University Press.

Tugby, M. (2022b). The laws of modality. *Philosophical Studies* 179: 2597–2618.

Vetter, B. (2015). *Potentiality: From Dispositions to Modality*. Oxford: Oxford University Press.

Vetter, B. (2018). Evolved powers, artefact powers, and dispositional explanations. *Aristotelian Society Supplementary Volume* 92: 277–297.

Vetter, B. (2021). Explanatory dispositionalism: What anti-Humeans should say. *Synthese* 199: 2051–2075.

Walsh, D. (2008). Teleology. In M. Ruse (Ed.), *The Oxford Handbook of Philosophy of Biology*, pp. 113–137. Oxford: Oxford University Press.

Walsh, D. M. and Ariew, A. (1996). A taxonomy of functions. *Canadian Journal of Philosophy* 26: 493–514.

Williams, N. E. (2019). *The Powers Metaphysic*. Oxford: Oxford University Press.

Wimsatt, W. C. (1972). Teleology and the logical structure of function statements. *Studies in History and Philosophy of Science* 3: 1–80.

Witt, C. (2003). *Ways of Being: Potentiality and Actuality in Aristotle's Metaphysics*. Ithaca: Cornell University Press.

Witt, C. (2008). Aristotelian powers. In R. Groff (Ed.), *Revitalizing Causality: Realism about Causality in Philosophy and Social Science*, pp. 129–138. Abingdon: Routledge.

Woodfield, A. (1976). *Teleology*. Cambridge: Cambridge University Press.

Wright, L. (1973). Functions. *Philosophical Review* 82: 139–168.

Acknowledgements

First, I would like to thank Durham University for granting the period of research leave during which this Element was written. Detailed and invaluable comments on drafts of the Element were kindly provided by Giacomo Giannini, Samuel Kimpton-Nye, and two anonymous referees. I am truly grateful to them, and also to the Series Editor, Tuomas Tahko, the Project Manager, Dhanalakshmi Narayanan, and the production team for excellent support throughout. I would also like to thank Ben Young for his excellent proofreading work. I received helpful feedback on the book's content from audiences at events in Dalhousie University, Durham University, the University of Bristol, and the University of Reading. Over the years, I have also benefitted from discussions with Greg Mason, Stephen Mumford, Simon Oliver, Ben Smart and many other colleagues about the themes in this Element. Much of this Element was written at The Hub in Consett, County Durham, and I would like to thank the wonderful staff there for their kindness and hospitality.

Finally, I owe a huge debt of gratitude to my family, and especially to Priya and Jayan for their patience whilst I was busy writing.

This Element is dedicated to Sapna Tugby.

Cambridge Elements

Metaphysics

Tuomas E. Tahko
University of Bristol

Tuomas E. Tahko is Professor of Metaphysics of Science at the University of Bristol, UK. Tahko specializes in contemporary analytic metaphysics, with an emphasis on methodological and epistemic issues: 'meta-metaphysics'. He also works at the interface of metaphysics and philosophy of science: 'metaphysics of science'. Tahko is the author of *Unity of Science* (Cambridge University Press, 2021, *Elements in Philosophy of Science*), *An Introduction to Metametaphysics* (Cambridge University Press, 2015) and editor of *Contemporary Aristotelian Metaphysics* (Cambridge University Press, 2012).

About the Series

This highly accessible series of Elements provides brief but comprehensive introductions to the most central topics in metaphysics. Many of the Elements also go into considerable depth, so the series will appeal to both students and academics. Some Elements bridge the gaps between metaphysics, philosophy of science, and epistemology.

Cambridge Elements

Metaphysics

Elements in the series

Laws of Nature
Tyler Hildebrand

Dispositions and Powers
Toby Friend and Samuel Kimpton-Nye

Modality
Sònia Roca Royes

Indeterminacy in the World
Alessandro Torza

Parts and Wholes: Spatial to Modal
Meg Wallace

Formal Ontology
Jani Hakkarainen, Markku Keinänen

Chemistry's Metaphysics
Vanessa A. Seifert

Ontological Categories
Katarina Perovic

Abstract Objects
David Liggins

Grounding, Fundamentality and Ultimate Explanations
Ricki Bliss

Metaphysics and the Sciences
Matteo Morganti

Teleology
Matthew Tugby

A full series listing is available at: www.cambridge.org/EMPH

www.ingramcontent.com/pod-product-compliance
Ingram Content Group UK Ltd.
Pitfield, Milton Keynes, MK11 3LW, UK
UKHW021948040125
452928UK00025B/672